THE SUNDAY TIMES

BUILDING YOUR OWN HOME

THE SUNDAY TIMES

BUILDING YOUR OWN HOME

EDITED BY CATHERINE MONK

In association with **Build It** magazine

hamlyn

Contributors

Edited by Catherine Monk

Chapter 1: Jane Crittenden and Kevin Smith

Chapter 2: Mike Dade and Roy Speer

Chapter 3: Peter Smithdale and Catherine Monk

Chapter 4: Nigel Grace

Chapter 5: Peter Smithdale

Chapter 6: Nigel Grace

Chapter 7: Laura Pank, Kevin Smith and Peter Smithdale

Chapter 8: Peter Smithdale

Chapter 9: Jane Crittenden

Chapter 10: *Build It* contributors

A NOTE ON PRICES

Prices are only given in British Pounds because the conversion rate between British Pounds and Euros fluctuates. All prices are accurate at the time of compilation but are liable to change.

First published in Great Britain in 2005 by
Hamlyn, a division of Octopus Publishing Group Ltd
2–4 Heron Quays, London E14 4JP

ISBN 0 600 61125 6
EAN 9780600611257

A CIP catalogue record for this book is available from the British Library

Printed and bound in China

10 9 8 7 6 5 4 3 2 1

Contents

Introduction

Do you have to be mad to self-build?

This is one of the most frequent questions I am asked as editor of *Build It* magazine and my answer is 'No'. In fact, people are probably mad not to self-build. Why? Because building your own home gives you the perfect opportunity to create the ideal house – the right size, in the right place, and to a design that you like.

Renovating a property or converting a building often means working within the constraints of an existing structure. This is, in itself, an exciting and creative challenge and one that can reap many rewards. But imagine having a blank canvas where you can outline the exact position of every room in a house to suit all your needs. You can also style it to suit your taste, instead of having to live with an existing design. These are the main attractions of a self-built property.

Around 20,000 people self-build in Britain each year and, given the right support, I believe there would be many more, which is why we have worked together as a team to produce this book. It covers everything there is to know about creating your dream house from scratch. Much of this information also applies to extending your property, or renovating or converting an existing one as well. The advice given is designed to help you with the project. However, as each home is unique, we strongly recommend that you also seek professional advice before embarking on a project.

WHERE TO START?

The first hurdle for anybody is raising the cash. These days, low bank interest rates and competition in the financial markets has made borrowing much easier for a greater number of people. Self-builders are often mistaken into thinking they don't have enough money to spend. Once it's pointed out that they can borrow money in much the same way as they do on an existing property, however, they soon realize that financing a self-build with a mortgage is actually quite straightforward.

The second challenge is finding the land. With an increased interest in the self-build sector and, more significantly, with the returns big property developers are enjoying on their developments, it is not unusual for available land to be bought almost as soon as it comes up for sale. Despite this, however, the more unusual plots or smaller pieces of land between houses tend to stay on the market for longer. These are usually less interesting to big property developers, and ideal for the self-builder.

REALIZING THE DREAM

The design of your house is something you can work on yourself or, if you prefer, consult a design professional. There are many things to consider

You do not have to confine your search for a good plot to areas of open land – for many self-builders the starting point is an existing house.

Building your own home offers you the chance to live in your ideal house. From the overall design and layout to bathroom fixtures and kitchen fittings, everything can be built to suit your every need.

about the structure of the building: do you want your house to be made from a timber-frame or the more conventional brick and block? What about the foundations you will need and the wall structures? Although this is the stage where you are most likely going to need to employ a professional, knowing exactly what is available will help you make an informed choice.

Inside the house, you have many more decisions to make regarding the heating, lighting, fixtures and fittings. Not only that, but with the latest developments in home automation, you can use music, home security and computer systems anywhere in the house at the press of a button.

Once actually building your house, you will undoubtedly be using the services of a number of professionals – a builder, plumber and electrician. This book will help you find and choose the right professional for the job, highlighting some of the typical pitfalls that could easily be avoided.

NO TASK TOO BIG

If any, or all, aspects of building your own home become daunting, don't forget that there is a number of design and package companies which have many years' experience of working with self-builders like you. They will happily guide you through any aspect of your house-building project, making the process as simple and stress-free as possible. With them, and this book, you should be well on the way to taking the self-build plunge in creating the house that you desire.

All the writers contributing to this book are experienced in their fields. Most are building professionals – architects, surveyors or planning consultants – while others are professional journalists who have accumulated knowledge through regularly writing about building issues for *Build It*. With many years' practice of renovating, I have been able to draw on my own experiences when editing this book. I hope you find it useful, and that it helps you to realize your dream – big or small.

Catherine Monk
Editor of *Build It* magazine and self-build expert for *The Sunday Times*.

Budget and finance

Where do I find land?

Calculating a budget for a building project is a task many of us would find extremely daunting. Yet a well-prepared and properly managed budget is critical for a successful project, otherwise you may be left with a half-built house and no money to finish it. The first thing you need to do is work out a rough estimate, which you can then develop into a realistic working budget that is appropriate for you and your project.

Establishing what you can afford

Always work out how much you can afford to spend before you do anything. Don't waste too much time on complex calculations at this stage: you just need to know what the figure is and how much further you can stretch it, if you need to.

To work out your maximum potential budget, answer the following questions:
- How much available cash do you have?
- How much equity do you have in an existing property or in stocks and shares?
- How much can you, or you and your partner, earn?
- How much money can you raise by borrowing from a bank or building society?

Often self-builders can be pleasantly surprised at how much money they can borrow from a bank or building society and at how relatively straightforward it is. They are also often surprised to learn that it is possible to borrow money on land and not just on a building. Securing a self-build mortgage is not that different from obtaining a conventional one on an existing property: you can go to a high-street lender or broker, who will be happy to help you with your borrowing requirements. As with most things, it is best to shop around to get the best deal.

There are many flexible self-build mortgage options, and it is important to get one that suits your personal requirements. One of the unique aspects of

It is essential to know the type of house you want to build right from the start as this may have a considerable effect on both budget and timescale.

SELF-BUILD MORTGAGE CASH FLOW

Money for a self-build mortgage is typically released in the following stages:

1 Procurement of the land.
2 Building of the foundations.
3 Construction of timber-frame, or wall-plate level for houses built in brick and block.
4 Completion of measures for being windproof and watertight.
5 First-fix (electrical wiring and plumbing, drainage, glazing, plastering).
6 Second-fix to completion (electricity and water on, tiling, decorating and final finishes).

Do not think that all land is available at the same price. The cost will be determined by its location among other factors, and this varies considerably across the country.

a specialist self-build mortgage is that money is released in stages as the project progresses. Typically there are between four and six stages (see left), and you can choose whether you prefer the payments to be made in advance of a stage or once it has been completed. As with conventional mortgages, you can elect to have a fixed or variable rate of interest or a self-certified mortgage. Clearly not all lenders offer the full range of variables, so it is worth talking to a specialist broker or reading the latest information in self-build magazines to ensure you get the right mortgage for your needs.

FOR LOVE OR PROFIT?
Before you embark on a project, you'll also need to have a clear idea as to whether you are building a house for yourself for the long term, or whether you want ultimately to make some money and move up the property ladder. This fundamental decision may dictate your expenditure throughout the build, from type of build to fixtures and fittings.

Essentially, if your intention is to develop land to make a profit, you are less likely to choose the more expensive options.

This does not mean that building your own home for the long term is without profit. The land-finding service, PlotSearch, discovered, when it collated some information on new builds, that self-build houses were often worth more than similar types of houses built by a developer (see above). A possible reason for this is that new builds are often of a more original design and are generally constructed using better-quality materials.

THE COST OF THE LAND
Arguably, the biggest factor affecting your budget calculations will be the cost of land. Traditionally, land costs amount to 20 to 25 per cent of a self-builder's available funds. However, with the rise in land prices, the average amount self-builders spend on land is more like 30 per cent of the total budget. In the main, land prices also reflect the current property market, so the south east, for example, is a lot more expensive than the most northern areas of Scotland. To work out what you can afford, take a look at listings included in many of the self-build

You can make significant savings during the build if you use less expensive materials – without affecting the potential sale price of the property on completion.

magazines available. To search for land, you could also sign up to specialist websites, such as the Landbank and PlotSearch, where you can look for the latest plots for sale in the regions in which you are interested.

THE SIZE OF THE HOUSE

Another considerable impact on your budget will be made by the size of house you want. Obviously, if you are determined to build a big house, then you will need a larger budget than most – or you will need to build in an area where land is cheaper so that you have more money to allocate to construction and materials.

There are several ways of working out the size of house needed to meet your requirements (see Chapter 3: Design, pages 56–69). For budgetary purposes you can get a good idea of the space you need by measuring up the size of your current living area. Establish the length of the house and multiply this by the width to produce the approximate ground floor area. Multiply this by the number of

It is tempting to exceed your needs when thinking about the kind of accommodation you want to include, but bear in mind this will be reflected in your costs.

> ### CALCULATING HOUSE SIZE
> All house measurements are typically described in terms of the internal dimensions, that is, where you walk around, as opposed to the outer shell. If you are taking the measurements from the outside, remember to deduct 30 cm for the construction of each of the external walls.

floors in the house to calculate, approximately, how much space you have. You can then use this figure as a basis for the size of your new house. For example, if your current accommodation is too small by, say 25 per cent, add that to the end result. Remember, this example only gives a rough guide for initial budgeting. When you look at planning your living area in detail, this will probably change.

THE COST OF BUILDING

Once you have calculated the maximum you can afford, you can work out how much you have to spend per square metre of the house. Simply deduct the cost of the land from your grand total and divide the remaining figure by the size of house you want.

The often-quoted industry ballpark figure is around £750 per square metre, so if you are close to this you have made a good start. This figure is calculated on a self-build of a modest nature, with materials sourced at competitive prices. If your figure is higher than the guide, you should be able to spend a bit more than most in chosen areas of the self-build, for example, on the fixtures and fittings. If your figure is lower by around £100 per square metre or so, then you will have to watch your budget very closely. Any lower than this, and you are better off building a smaller house to get the cost per square metre to a reasonable level.

Although useful for providing a rough idea of costs, this is still only a guide. Actual costs, such as labour, can vary quite dramatically, as can land prices. For example, it is not unusual for a plumber in London to charge around £200 a day, while one in a sparsely populated region of Scotland may charge half that. Therefore it is always best to do your own research before embarking on a self-build project to make sure you can afford to build a house to your required standards and in your chosen location.

COST OF BUILDING A TIMBER-FRAME HOUSE OF A MEDIUM SPECIFICATION (2004 FIGURES BASED ON A 165 M² BUILDING)

Region	Self-managed	Project-managed
London & M25 Area	£205,733.04	£225,406.34
Midlands	£163,575.96	£179,033.56
Wales, Yorkshire, the North & Scotland	£149,523.60	£163,575.96
The Highlands	£140,155.36	£153,270.90

Data supplied by Insider Project Management

SEEKING PROFESSIONAL HELP

For a more accurate idea as to how far your budget will stretch, ask a project manager or quantity surveyor to calculate it for you. This usually involves him or her carrying out a financial feasibility study, for which you will be asked to clarify the region (to determine labour costs), size and type of house, and construction method. Both professionals will be able to give you costs that compare prices whether you self-manage or hire a project manager (see above).

EASY DOES IT

Don't worry if working out the budget in detail is not for you. A far simpler approach is to choose a design from a package company. Most design and package companies will provide a clear breakdown of costs, including the regional variations, of their standard designs. Many will also provide you with a project manager or a list of potential candidates familiar with their house designs. The advantage of this approach is that it is far less complicated for you. You also don't necessarily have to choose a standard design, as most can be modified. Other companies will help you plan your dream home with an in-house team of designers, while most package companies can also help you to find land and labour.

Planning your costs

The next step is to turn your budget into a cost plan. This is a list of the elemental stages of the build with approximate figures against each one. It will give you a more realistic idea of whether you can achieve what you want to build with the money you have to spend. The best time to do this is after you've bought a plot of land and have had the initial design prepared, but before you seek planning approval. There are no hard-and-fast rules, but if you do the cost plan too early you could waste time and money if your planning application is turned down.

The best way to start is by breaking each stage of the build into sections. Using a spreadsheet, list items in the order in which you would build them. First on your list will be the price of the plot and any surveys that might need doing. You will end with a budget for external works. In between should be stages like the services (water, gas, electrics), earthworks and drainage, foundations, walls, roof, windows and doors and so on (see page 13).

Compiling the list is the easy part of the cost plan, now you have to figure out what budget to attach to each stage. There are various ways you can get this information, but don't get too concerned: the cost plan is just the first stage in establishing your final estimate and, eventually, the actual costs of the build.

See the pie chart overleaf for a guide to the percentages usually allocated to the various stages of a build. As quotes and prices come in, you can update your original cost plan with the real percentages. This comparison will help you to see where you might under- or overspend. The stages will vary depending on your individual project.

ESTIMATED PERCENTAGE COST OF EACH STAGE OF A SELF-BUILD

Source: www.costofdiy.com

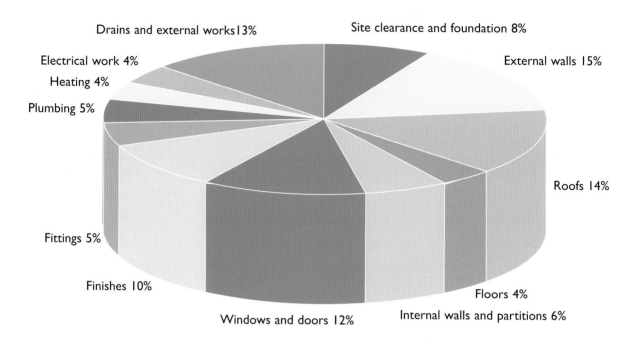

Drains and external works 13%

Site clearance and foundation 8%

Electrical work 4%

Heating 4%

External walls 15%

Plumbing 5%

Roofs 14%

Fittings 5%

Finishes 10%

Floors 4%

Windows and doors 12%

Internal walls and partitions 6%

COLLATING ESTIMATES

You will get the most accurate estimates by hiring the services of a professional – a quantity surveyor, project manager, building contractor or your architect – but there will be a cost for this. The more information you can provide early on, the more accurate the estimate will be: at the very least you should provide the house design, a brief specification on the proposed method of construction and the standard of finish. To save on fees, you can get a rough estimate of labour costs simply by ringing a few tradespeople. You'll need to talk to a plumber, a carpenter and an electrician. Ask each to give you a quote for his or her average daily rate and find out roughly how many days it would take to install a kitchen/fit a bathroom/wire a house, and so on. Remember that the more work you need a tradesperson to do in one go, the more likely you will be able to get a better discounted rate.

Confirming costs

Once the cost plan has established that you can build the size and style of house you want on the plot of land you have bought, you need to get the estimate quoted and the costs confirmed.

BILL OF QUANTITIES (BOQ) AND MATERIALS SCHEDULE

These are the two methods for getting a building project costed accurately.

The Bill of Quantities (BoQ) is drawn up from the information in your construction plan. It is a description of the materials, labour and plant required for each stage, listed in detail by quantity and unit price. The BoQ is used in tendering to provide a uniform description of the project for building contractors to quote on and makes it easier for you to compare costs later on.

The Materials Schedule is a list of the materials derived from the BoQ. It details the supply of materials only and does not include labour and plant costs. It shows every single item needed to complete each stage of the build and is effectively a 'shopping list' for ordering in supplies. You can give this to subcontractors and builders' merchants for quoting but you will pay a premium for their services, which would be free if you were to do it yourself.

EXAMPLE BUDGET FORECAST

The following tables provide examples of the budgeted cost of a house constructed of brick and block, of a good to average specification with modest labour costs. The final cost depends on the overall involvement of the self-builder.

COMPARISON OF COSTS DEPENDING UPON A SELF-BUILDER'S INVOLVEMENT

	Managing the project yourself	Managing it with a project co-ordinator	Managing it with a builder/ contractor
House build cost	£271,320.00	£298,452.00	£339,150.00
External works (add £2–10k)	£2,000.00	£2,000.00	£2,000.00
Garage (add £0, 7, 12k)	£0.00	£0.00	£0.00
Budget total	**£273,320.00**	**£300,452.00**	**£341,150.00**

HOW THE COSTS ARE SPREAD OVER EACH BUILD STAGE

Build stages	Mangaging the project yourself	Managing it with a project co-ordinator	Managing it with a builder/ contractor
Foundations	£40,998.00	£45,067.80	£51,172.50
Wallplate	£54,664.00	£60,090.40	£68,230.00
Roof/first-fix	£54,664.00	£60,090.40	£68,230.00
Plaster/insulate	£40,998.00	£45,067.80	£51,172.50
Fitted out	£81,996.00	£90,135.60	£102,345.00
Total	**£273,320.00**	**£300,452.00**	**£341,150.00**

NB The percentage of the total cost due at each stage may vary according to build size and type.

There are four ways to get costs confirmed:
• using a building contractor
• using a project manager
• managing the project yourself
• employing a quantity surveyor.

There is no right or wrong option – it just depends on your project. For example, the first route passes the risk on to the subcontractors, which might be a better option for a more complicated house design.

METHOD 1: USING A BUILDING CONTRACTOR

A building contractor will build the structure of your house and then hire subcontractors to do the electrics, carpentry, plumbing etc. It is likely that your building contractor would buy in materials because he or she will have a good relationship with local builders' merchants and hire subcontractors on a daily rate for fix only. He or she will add a percentage on for the project management role and a percentage for the materials and plant.

METHOD 2: USING A PROJECT MANAGER

A project manager will price your house build according to your plans as part of his or her services in managing the project. There are a number of ways he or she can do this:

1 Use a BoQ for the builders and subcontractors and ask them to provide a cost including labour, plant and materials.
2 Pass a Materials' Schedule on to the builders' merchant (either prepared by you or him or her) and then give a BoQ to the subcontractors requesting a cost for fix only.
3 As above (no.2) but hiring tradespeople on a daily rate.

METHOD 3: SELF PROJECT MANAGEMENT

You'll need to obtain a detailed technical specification of your house from your architect (and this is still relevant if you hire a project manager or building contractor). But make sure you also have a good discussion with your architect so that you understand exactly how he or she envisages your house will be built. He or she will have certain ideas about how parts of the build will be done and what kind of materials will be used, and he or she needs

to convey all of this to you. Make sure that these conversations are put in writing.

You'll need to decide what level of involvement you are going to have as a project manager. If you hire the builder and the subcontractors, as well as buying in the materials, you will need to produce a BoQ for fix only and a Materials' Schedule for the builders' merchant. If the subcontractors are going to source their own materials, however, you need to provide them with a BoQ and request a price for labour, plant and materials. If the builder is going to hire the subcontractors on your behalf, he'll do his own Materials' Schedule from your BoQ.

METHOD 4: EMPLOYING A QUANTITY SURVEYOR

Spending a small amount of money to hire a quantity surveyor or project manager to do a BoQ and/or Materials' Schedule for you has real benefits. He or she will do it much quicker and you can be assured it will be accurate. For example, some quantity surveyors have their own software for drawing up a labour-and-material cost analysis for the self-builder, in which the project is broken down into a number of main stages with every major building material quantified and priced from a product database. A schedule is also put together, which shows the start and end dates of each stage of the build.

If you follow this route, you are likely to have quite a detailed discussion with the quantity surveyor, listing the various elements of the build: what kind of kitchen you want fitted, for example, right down to floor and wall finishes, and so on. From this information he or she will be able to provide you with detailed building costs based on certain assumptions about the house size, the house style and the regional labour costs.

Pricing is one of the fundamental roles in a quantity surveyor's job and he or she can provide you with a BoQ and Materials Schedule based on the industry-wide Standard Method of Measurement (SMM7). This examines each element of a build in great detail, calculating size and materials in order to reach an accurate price.

ESTIMATING SOFTWARE

For a few hundred pounds, estimating software can make things less complicated, because the elemental breakdown is already set up in the package. All you'll need to do is enter the information from your plans and the calculations will be done for you. For example, if you enter the dimensions of the floor, the software will list the materials and work out the quantities. The programme sets out the stages in the order of the build, but materials are also grouped together separately so that you can send the information to a builders' merchant for costing.

A typical software package will also update the spreadsheets as and when you change prices and will provide a format suitable for printing. Some software packages are affiliated to a builders' merchant, and provide regular price downloads to ensure you are kept abreast of any changes. Try EasyPrice Pro, HBXL's Estimator & Project Manager, and Fast Estimate.

SENDING YOUR PROPOSAL OUT TO TENDER

When it comes to tendering the job, whether it be from a builder, subcontractor or builders' merchant, it is always best practice to get three different quotes. This allows you to review the costs for consistency, and will give you a market-tested price that reflects current trends and, therefore, value for money. Check the costs against your cost plan and you will be able to see if there are any major differences from your own estimates. Watch out for companies who only price labour and plant and insist on you providing materials, as this will distort your comparisons.

Once you have decided which tender to accept, it is sensible to get the data loaded on to your computer. How you decide to do this is a matter of preference. If you're using estimating software there will be a template provided for this anyway. But if you are setting up your own spreadsheet, you could add another column to your cost plan and input the quotes, or expand on this by using the more detailed information provided in the BoQ.

Monitoring the working costs

There is no point making all the cost preparations if you forget to keep an eye on the costs during the build. This doesn't have to be a daily chore but it is a good idea to update them on your spreadsheet every week or fortnight, depending on the size of your project. Keep a copy of the original – either in the form of a separate spreadsheet or as a hard copy – before you update it. That way you'll have a record of the changes throughout the project.

Be aware that, no matter how well defined the costs are, there will always be some element of change simply because you cannot predict every scenario. For example, the council's building inspector might not be happy with the ground conditions when it comes to approving this stage, and may tell you to build a different type of foundation. Bad weather can slow progress and materials can get accidentally damaged. Even as you build, you might change your mind about something, which can also affect the budget. All of these

COSTS FOR RENOVATIONS AND CONVERSIONS

It is much harder to get accurate costs for renovations and conversions, simply because you do not know what you might find until you start doing the work. The best way to minimize enormous changes to the budget is by allocating money to pay for professional advice. Get a structural engineer to do a detailed structural survey; work closely with the council's planning department and building controller; and consider employing a project manager with experience in your kind of project. Above all, allow the contingency to be at least 20 per cent of the total building costs.

Unforeseen circumstances and bad weather conditions can slow progress: even a day's delay can increase bills for labour and materials.

reasons confirm the need for a realistic contingency (see box, page 15). It is essential that you know of any price changes as soon as possible before the work is carried out. Ideally, variations/alterations should be put in writing before anything happens so you have confirmation that the decision was made and something to refer to when the invoices come in. This isn't always possible immediately, but you must make sure you do it as soon as it is practical. Don't wait until the job is finished and don't forget to update your spreadsheet with the new quote.

Reclaiming value-added tax (VAT)

One of the main advantages of building your own home, or converting a non-residential dwelling, is that you can save money by reducing the amount of value-added tax (VAT) you incur on the project. This is because HM Customs and Excise don't currently charge VAT on bare building land and many building services and materials used to create new homes. This applies to property developers, mass house builders and, of course, self-builders. (This situation was correct at the time of writing. *The Barker Review*, published in conjunction with the 2004 spring budget, indicated that zero-rated VAT on new builds may be abolished in the future.)

Unique to self-builders is a refund scheme that guarantees any VAT incurred (when buying goods) can be easily refunded on completion of the project. Overall savings can be between 5 and 10 per cent of the cost of a self-builder's project.

Although this is good news, the self-build VAT reclaim process can be a little complicated, particularly when defining the elements of a build that do or do not qualify for a refund. Use this section to guide you through the process. However, as these facts may change with time, it is also advisable to find out what the current status is directly from HM Customs and Excise when you decide to build.

CAN ANYONE CLAIM BACK VAT THROUGH THE SCHEME?

Anyone who is building a home, or converting a non-residential building that hasn't been lived in for ten years or more, and which is destined to be his or her main residence, is entitled to use the VAT refund scheme. However, there are instances when the scheme cannot be used:

- The property is being built to rent out (the builder never intends to live in the new building and it won't be his or her new home);
- The property is being built to sell on immediately (the builder is being a speculative developer);
- The property is destined to become a bed and breakfast;
- The property is to become a care home, charging fees to residents;
- The property is to become to home of a membership club or association.

ON WHICH GOODS CAN I CLAIM BACK THE VAT I INCUR?

In simple terms you can submit a claim for any goods that become permanently 'incorporated' into the house, that is, you couldn't take them with you if you decided to move. These include:
- hard building materials (bricks, cement)
- doors
- windows
- curtain poles and rails
- decorating materials (emulsion and wallpaper)
- flooring (wooden or tiles, but not carpet)
- fitted kitchens (not conventional cookers and hobs)
- solid fuel cookers
- fireplaces

- fire alarms
- smoke detectors
- boilers
- water tanks
- plumbed in appliances (baths, toilets and showers)
- lifts
- light fittings
- TV aerials
- wiring
- air-conditioning units
- burglar alarms
- built-in vacuum systems
- permanent boundary fencing
- soft landscaping (turf and topsoil).

Be careful with soft landscaping, as you will only be able to claim for anything that was detailed on your original planning drawings.

Also, remember that you are obliged by law to carry out everything detailed on your planning drawings so it may not pay in the long run to draw on more trees than you intend to plant, just because you will be able to claim back the VAT you incur.

You can submit a claim for any goods that you buy and give to your builder or that your builder buys on your behalf.

As long as you intend to live in the house you are building you can claim back the VAT on all building materials and many internal fittings.

WHAT CAN'T I CLAIM THE VAT BACK ON?

Basically, you can't claim the VAT back on anything that isn't permanently fixed to a house, that is, anything you could take with you if you decided to move. This includes:
- fitted furniture (except kitchens)
- unfitted furniture (sofas)
- cookers (except a range fuelling a central-heating system)
- washing machines
- tumble dryers
- fridges
- freezers
- waste-disposal units
- doorbells
- carpets
- underlay
- carpet tiles
- garden ornaments
- sheds
- greenhouses
- tools
- equipment
- consumables (sandpaper and white spirit).

The above lists are not comprehensive, but should give you some idea as to what you can claim the VAT back on. For example, white goods (fridges and washing machines) are often incorrectly included in submitted reclaims, which can result in frustrating delays. If you have any doubts you can call the Customs and Excise advice service for information.

WHAT ABOUT LAND?

The purchase of land does not qualify for the VAT refund scheme. However, you should not pay VAT on bare building land (with no service connections): just let the owner of your plot know that you intend to build your own home on the site. Furthermore, if you are purchasing a single plot with outline planning permission, most parties involved in the sale will be aware that VAT shouldn't be charged. Be careful here: the seller may be unaware that you intend to build a home and might honestly charge you VAT, which you won't be able to claim back from HM Customs and Excise when you submit your claim on goods.

If you buy a plot with service connections in place, you may encounter some legitimate VAT, but this won't be charged on the whole value of the site. Instead, VAT will be charged only on the proportional value of the utility infrastructure, and the seller will be responsible for working out what proportion of the land is liable.

CAN I CLAIM BACK VAT ON A CONVERSION?

As with land, VAT will not be added to the purchase price of a non-residential property, for example a barn or a water tower, if you say before you buy it that you are going to turn it into a home. Again, it is essential you make you intentions known from the outset: if you pay VAT in error, you will not get another chance to claim it back. As with building a new house, you will also be entitled to claim back any VAT incurred on goods, but at a reduced rate of 5 per cent rather than 0 per cent.

HOW CAN I AVOID PAYING VAT ON SERVICES?

Services do not qualify for the refund scheme. However, if you are constructing an entirely new home, builders and contractors should zero-rate their services, which means you don't pay any VAT from the outset. The situation is slightly different with conversions, where builders and any relevant contractors are entitled to charge a reduced VAT rate of 5 per cent. Again, make sure you are paying the right amount of VAT, as you will not be able to apply for a refund through the scheme.

HOW ABOUT RENOVATIONS?

Under normal circumstances, you will not be able to make a claim for VAT incurred on goods used in a renovation unless you can prove the building has been uninhabited for more than ten years. The same applies to builders' services – if the property has been inhabited during the past ten years you will incur VAT. (The ten-year rule is measured from when the work starts and not from when you move in.) To prove that a property has been empty for the past ten years you will have to speak to an empty property officer at your local authority. He or she will be able to write a letter confirming that the building has been uninhabited, and this will be enough to support a VAT reclaim claim. If the officer does not have records of the property's inhabitants, you could try researching old electoral rolls, looking at council tax data or speaking to the utility companies to see if they have any records.

WHAT ABOUT EXTENSIONS?

You will not be entitled to zero-rated VAT if you are building an extension on to an existing property. Naturally this also applies to conservatories, patios, garages and other improvements made to an already completed home. It is worth bearing this in mind when you are at the planning stage. For example, if you wait until you have finished building your home before adding a conservatory, you will miss out on getting it at a VAT-free price.

You can claim back VAT encountered on goods, and qualify for zero-rated services, if you buy an incomplete house (just the foundations or the shell) and finish it yourself. However, you cannot reclaim VAT incurred on modifications to a brand new house built by a developer.

WHAT ABOUT VAT ENCOUNTERED WHEN PURCHASING GOODS?

The first thing you need to do is get in contact with the Customs and Excise National Advice Service and ask them to send you a copy of Notice 719 (VAT refunds for 'do-it-yourself' builders and converters). This handy booklet is an official one-stop guide and will help you fully understand the refund process. Ask them for a claim pack or visit their website for the relevant forms that you will need to complete.

It is rare to qualify for a VAT reclaim on an extension to an existing building, unless it has been uninhabited for ten years or was incomplete at the time of purchase.

You can submit a claim as soon as your project is finished, but no later than three months from the house or conversion being completed. You will need the relevant completed forms and supporting evidence. This will include credit notes, bills, invoices, a copy of the planning permission and plans of the building. You will also need proof that the project is complete – either a certificate of completion from your local authority, a letter from your local authority, a council tax assessment or a certificate from your bank declaring the project complete.

A successful claim obviously needs a lot of care and attention so make sure you maintain good financial records throughout the building project. Keep all paperwork in a safe place and in a sensible order. If you are successful in this you will have no problems with your refund.

WHEN WILL I RECEIVE THE MONEY?

Provided your claim form has been completed properly, and you attach all the supporting evidence, you should receive your refund within 30 working days.

Structural warranties and site insurance

You will be spending a considerable amount of money on your self-build, renovation or conversion, and it makes sense to protect your investment. There are two principal ways of doing this: the first is to take out a structural warranty on the building; the second is to take out an insurance policy that covers the building site and anyone who comes into contact with it. It is important to note that, although the two documents may seem similar, they actually cover very different things.

STRUCTURAL WARRANTIES

A structural warranty works rather like a guarantee for an electrical appliance, such as a television. It is a certificate that proves your product, in this case your building, has been completed to the required standards. It also covers you for any necessary repairs or replacements should the project develop problems related to its structure. The cover usually runs for a period of ten years after the house is built, although the length of time depends on the warranty provider.

The warranty will be issued by a specialist provider, such as the National House-Building Council (NHBC) or Zurich Municipal. The provider may make regular inspections throughout the build (between five and eight times) to make sure that everything is built to the required standard. Alternatively the provider may rely on the standard inspections made by the local authority's building control department. Following a final visit, providing everything is completed satisfactorily, the provider will issue a structural warranty certificate. A warranty naturally comes at a cost – usually a one-off set fee – which varies depending on the provider.

You can expect a number of site visits by a building control officer, who needs to inspect key stages of the build to make sure you comply with Building Regulations.

Do I really need a structural warranty?

You are under no legal obligation to take out a structural warranty. A building control officer will also be visiting your house at certain stages during the build. Furthermore, your project may have a degree of cover provided by the personal indemnity insurance (PII) that your architect, architectural technologist, surveyor, builder or other contractor should have.

However, these forms of 'cover' are not as reassuring for the self-builder, neither do they last as long. Building control inspections only guarantee that the building is of the required standard at the time of completion. They do not provide a guarantee for repairs or replacements that may need to be carried out following completion. If anything happens at a later date and you don't have a structural warranty, the chances are you'll have to pay more money to rectify the problem.

Personal indemnity insurance also only lasts for as long as the professional keeps the policy up to date. If his or her personal cover runs out, your building no longer has any protection and you are not in a position to make any kind of claim if things start to go wrong. Also, if you decide to sell your home, in most cases this kind of cover cannot be transferred to the new owners. Furthermore the PII only covers the work that the professional oversees. This means that you cannot claim negligence for work a builder does, say, on an architect's policy. This is because the policy is aimed to protect the policy holder and not the self-builder. To make a claim, the self-builder would have to prove that the professional with the PII was negligent and this could be difficult, particularly if he or she wasn't on site at the time the work was done.

Warranties are also important when it comes to selling a house. If your buyer needs a mortgage to fund his or her purchase, it is becoming increasingly likely for the lender to ask for proof that the property has a structural warranty before releasing any funds to the borrower. Even if a buyer does not need a mortgage, he or she might be put off if the property doesn't have a warranty, on the grounds that buyers generally expect new houses to have them.

It is worth investing in a structural warranty if you intend to build using a non-conventional construction method (timber or steel frame, for example). This is because, quite often, prospective purchasers will be unfamiliar with the building method and may have doubts about the quality and durability of the construction. The same applies if you want to incorporate unusual, design-led features in the home. A warranty should be all you need to convince a buyer that the building is structurally sound.

When do I need to think about obtaining a warranty?

You should look into a structural warranty as soon as building work begins, as many of the providers insist on thorough inspections throughout the construction process in order to issue the certificate. Many work in conjunction with the building control department, who will also make inspections as the build progresses. It is worth bearing in mind that you may not be able to obtain a warranty once your building is complete, should you suddenly decide you want one.

What exactly does a structural warranty cover?

You should always go through a policy with the provider to determine exactly what cover you are buying. Make sure you check any small print on the policy forms. At the simplest level the following elements should be included:

- Protection if your builder goes bankrupt during the build.
- A guarantee that your builder will rectify any problems that occur in the first two years after completing the work. It is common for the provider to check that the builder is reputable before the cover begins.
- Reimbursement of any repair to the structure of the building that is necessary within the remainder of the policy term. The repairs usually have to cost more than a specified amount, which will be stipulated in the policy.

Some policies incorporate the following elements or, for an additional fee, you can probably have a warranty that extends to these areas:

- Special policies aimed at self-builders doing most of the work themselves or taking responsibility for project management.

- Policies offering clean-up cover should you discover your site is polluted or contaminated.

Besides NHBC and Zurich Municipal, there are a few smaller providers offering structural warranties underwritten by a large insurer. As with any company, you should investigate their credentials and suitability to your needs and always get references from previous customers. It is worth considering the likelihood of the company still being in existence in ten years' time.

SITE INSURANCE

In the simplest terms, site insurance offers protection for your building site and anyone visiting or working on it while construction is in process. The exact terms of the agreement will depend on both the policy and the provider. Site insurance usually lasts for a set term – usually a couple of years – which should cover the time of the build.

What does site insurance cover?

Site insurance can cover a variety of areas, so make sure you are happy that the policy includes everything you need. As with all insurance policies, the level of cover depends on the provider. When you are getting more than one quote from a company, make sure you look at comparable offers. Most site insurance packages for self-builders cover these three main areas: public liability insurance, employer's liability insurance and site cover.

1 **Public liability insurance** This insurance provides cover in the unlikelihood that any member of the public incurs an injury while on your building site, whether visiting or working on the site. Public liability insurance can also provide cover should any accidental damage happen to another person's property during the construction of your house, caused by a worker on your site. For example, a delivery lorry may accidentally knock down your neighbour's wall.

    If you are employing a building contractor, it may not be necessary for you to take out public liability insurance as they may have their own policy. Be cautious, however, and make sure all of the paperwork is up to date. It is worth asking to see a copy of the policy to know exactly what you

You can expect your building contractor to have his or her own public liability insurance. The terms vary, however, so make sure it provides the cover you need.

are, and are not, covered for. If a member of the public does sustain an injury resulting from work carried out on your site, and your contractor's policy proves invalid, you will be held responsible and the cost could be quite considerable. Even if your builder does have up-to-date cover, it would probably be worth investing in a policy of your own, just to be on the safe side.

2 Employer's liability insurance Very few self-builders complete every part of a project themselves. In most cases, you are likely to be employing builders and specialist subcontractors to assist the construction process. While these people are working on site, you are their employer and you need to cover yourself should one of them sustain an injury. The same applies to friends and family who help you out.

Public, and employer's, liability insurance are practically essential in today's society for countering any potentially costly compensation claim. If you have insurance of this kind, at least you know that you are covered in the event of an accident.

3 Site cover Site cover is virtually the same as house insurance: it literally provides protection for your building plot and its contents, and usually covers the following:
• theft of materials
• site vandalism
• storm and flood damage

• fire damage
• accidental damage incurred during the construction process.

While site insurance is not essential, you should consider what you have to lose if you don't invest in a policy. If you are on a tight budget, the theft of materials might be a disaster. Also, consider that expensive kitchen appliances, bathroom fittings and fires could also be stolen if they are delivered before they can be fitted – especially if your site isn't under constant surveillance at night.

Other areas that you may want covered by your site insurance policy, or separately with another policy, are:
• Personal injury: If you are going to be working on site yourself, it is worthwhile investing in an extension to a standard policy. Most providers have the facility to add personal cover to a policy – probably for an additional cost.
• Caravan or temporary building: You may want extra cover in case somebody vandalizes a temporary building on your site.

There are many potential hazards on a building site and entering into a new-build project without some form of site cover and/or public liability insurance is unadvisable.

- Existing structure: If you are converting a barn or renovating a building, you may want to have a policy that covers the existing structure as well.
- Legal expenses: Some policies can cover any legal dispute you may have with a contractor or supplier.

Who will provide site insurance?

There are a number of companies offering public and employer's liability cover and site insurance. Shop around for a good deal and choose one that explains things clearly and that you feel comfortable with. With some of these providers the cover converts to buildings insurance when a project is complete. These are usually specifically targeted at the self-build market and the cover extends to the end of the term of the policy, should you finish the build before your site insurance expires.

Many self-builders choose to rent temporary accommodation or live on site during the construction of their house.

Capital gains tax

Those people interested in making a profit through self-building need to know whether they are liable to pay any extra taxes, typically in the form of capital gains tax (CGT).

WHAT EXACTLY IS CGT?

This is a tax you have to pay if you dispose of (sell or give away) an asset that has increased in value since it originally came into your possession. An asset can be anything from shares to land and buildings and, with self-build, it could be your completed project. It is worth remembering that it is the *increase* in value of the asset that is taxed, not the full amount you receive as a result of the sale.

HOW IS CGT CALCULATED?

1 The gain on any assets that you have disposed of in the previous tax year is calculated.

2 If the total of your net gains in a tax year is less than a certain amount, called the annual exempt amount (AEA), you will not have to pay CGT.

3 If your net gains are more than the AEA you have to declare this to the Inland Revenue and they will work out how much CGT you have to pay. They do this by taking into account the rate of income tax you usually pay.

WILL I HAVE TO PAY CGT IF I SELL THE HOME I HAVE BUILT?

Under normal circumstances you shouldn't have to worry about CGT if you decide to sell your self-built home. A home is viewed differently from other assets and is entitled to relief against CGT. You should be entitled to relief in the following instances:

- the home was your only/main residence throughout the period you owned it.
- you used the property as your home all the time you owned it.
- your house doesn't exceed 5,000 m^2.

WHAT HAPPENS IF I NEVER LIVE IN MY NEW HOME?

Whether you have to pay CGT in this situation is largely dependent on whether you own another house at the time of sale. If you do (perhaps you chose to live in your existing home while you were self-building) the likelihood is you will have to pay CGT if you dispose of your self-build, because it will be difficult to prove the new building was your main residence.

If you own no other residential properties at the time you sell your self-build (perhaps you are living in rented accommodation or a caravan) you may be entitled to relief against CGT, as you will be able to prove the building is the only residential property you own.

WHAT IF I BUILD A SECOND HOME AND DECIDE TO SELL IT?

If you dispose of a second home it is likely that you will have to pay CGT on any gain in value, because a second property is viewed simply as an asset and not your main home.

WHAT HAPPENS IF I BECOME A SERIAL SELF-BUILDER?

If you intend to become a serial self-builder you should also be entitled to relief against CGT, providing you actually live in each of your houses (and each is your main home) for a period of time before you sell. This is usually considered to be at least one year. However, you may draw the attention of the Inland Revenue after building and selling a series of houses in quick succession, and they might question whether you have profit-making intentions.

WHAT HAPPENS IF I VENTURE INTO PROPERTY DEVELOPMENT?

You are entering new territory if you start building and renovating properties as part of a business enterprise, and CGT won't be an issue. Although you will undoubtedly have to pay a different tax on any profit you make, this will not be in the form of CGT and is more likely to be calculated when you record your profits – in an annual tax return, for example.

FINDING OUT MORE

Although CGT won't be of concern to most self-builders, is largely dependent on individual circumstances and is difficult to set down criteria that apply to everyone. If you think your project might be affected, it is best to get in contact with your local tax office or visit the Inland Revenue website for more information.

Costs incurred when purchasing land or property

There are certain fees and duties that you have to pay when you purchase any land or property, whether it is a new build, a renovation or a conversion. If you own a property that you need to sell before you can embark on your self-build project, there are additional fees to account for in your budgeting.

If you definitely intend to self-build for profit, it is advisable to take advice on CGT before beginning your project.

COSTS ON AN EXISTING PROPERTY

1 **Estate agent's fees** You will have to pay an estate agent (if you use one) the agreed fee for marketing and selling your property. This is usually a percentage of the sale price of the property.

2 **Solicitor's fees** A solicitor will charge a fee for arranging the sale of the existing property. This is usually a flat fee for the work, plus expenses. On rare occasions, a solicitor may negotiate on a percentage basis.

3 **Homebuyer's report** The Government has been paving the way for property owners to arrange and pay for a surveyor's report to be carried on the property for sale. If you do this, rather than the buyer, then you should include this charge as part of the expense of selling your existing property.

4 **Mortgage redemption fee** If you are locked into a mortgage for a specified time, the lender will probably charge a redemption penalty for permitting you to break the agreement before that time is up. There may also be a deeds-reduction fee added on to the mortgage account, which will also have to be paid.

COSTS OF BUYING A PROPERTY OR LAND

1 **Solicitor's fees** A solicitor is required to supervise the conveyancing process – the exchange of contracts between the seller and buyer – and to ensure that all parties concerned get the agreed price. For helping to buy the land or property and checking the details of any deeds on the land or property, a solicitor will charge a fee. This is usually a flat fee for the work, plus expenses. On rare occasions, a solicitor may negotiate on a percentage basis. Typically you will have the same solicitor working on the sale of your existing property and the purchase of the new property/land.

2 **Surveyor's report** A surveyor will make sure that the property or land you are buying is of an adequate standard for your intentions, and is at a reasonable selling price. In the case of an existing property you will probably choose between a homebuyer's report or, more likely, a surveyor's

report. In the case of land you will probably have a surveyor check that the land is suitable for building on, that it can take the weight of foundations, and that it doesn't have contaminated soil or is in a flood plain.

3 **Area searches** Initially, at the time of purchase, your solicitor will arrange for any searches of the area surrounding your site or property to be carried out. In theory you can do this yourself, but it is usually advisable to leave it to a solicitor, who knows the procedure in detail and is less likely to make mistakes. The searches highlight any future development close to the site that might put you off buying it. For example, you might not want to buy a piece of land if you discover there are plans to build a new motorway nearby in a few years' time.

4 **Additional reports or search fees** If any issues are raised by the surveyor, you will incur extra costs for additional reports or searches. These can often be arranged through your solicitor, or independently. For example, if a barn you are intending to buy does not appear to be sound, a surveyor will recommend an investigation by a structural engineer. In the case of buying land, it is quite common to have soil surveys carried out to check for levels of contamination.

5 **Stamp duty land tax** Anybody buying land or property in the United Kingdom is required to pay the Government a tax as part of the transaction – this is known as stamp duty land tax. It is payable at the time of exchange or completion and is calculated as a percentage dependent on the value of the land or building that you are buying (see box, page 26).

6 **Land registry fees** When ownership on a property or piece of land changes, you have to pay a fee. This depends on the value of the transaction (see box, page 26).

7 **Arrangement fees (for lenders)** If you are taking out a mortgage to fund your project, the lender may charge an arrangement fee. This is a fee you may have to pay up front (around £500) before any funds are released and the sale gets underway.

STAMP DUTY LAND TAX IN THE REPUBLIC OF IRELAND

Stamp duty is a tax the Government levies on the purchase of all second-hand residential property over the value of €127,001 and on all new property bought for investment purposes. New houses are exempt provided the property is for owner occupation and the floor area is less than 125 m².

Value	First time buyer	Other owner occupier
Up to €127,000 (£86,232)	0%	0%
€127,001 to €190,500 (£129,359)	0%	3%
€190,501 to €254,000 (£172,479)	3%	4%
€254,001 to €317,500 (£215,610)	3.75%	5%
€317,501 to €381,000 (£258,732)	4.5%	6%
€381,001 to €635,000 (£431,245)	7.5%	7.5%
€635,001 (£431,245) and over	9%	9%

Note that the applicable rate is charged on the full purchase price, not just the portion of the price above the threshold. For example, on a €200,000 house, you pay the €200,000 rate on the entire purchase price.

New properties bought by a first time buyer/owner occupier measuring more than 125 m² will be liable for stamp duty calculated on either the site cost or 25 per cent of the total cost (site cost plus building costs), whichever is the greater figure.

The following are the rates of stamp duty which apply to sites and land:

Value	Rate of stamp duty
up to €10,000 (£6790)	0%
€10,001 to €20,000 (£13,580)	1%
€20,001 to €30,000 (£20,377)	2%
€30,001 to €40,000 (£27,169)	3%
€40,001 to €70,000 (£47,546)	4%
€70,001 to €80,000 (£54,331)	5%
€80,001 to €100,000 (£67,914)	6%
€100,001 to €120,000 (£81,497)	7%
€120,001 to €150,000 (£101,850)	8%
€151,001 (£101,850) and over	9%

Rate of exchange accurate at time of compilation.

STAMP DUTY LAND TAX IN UK

The percentages in England, Scotland, Wales and Northern Ireland are currently as follows:

Value	Rate of stamp duty
£0–£60,000	Exempt
£60,001–£250,00	1%
£250,001–£500,000	3%
£500,001 and above	4%

Note that the applicable rate is charged on the full purchase price, not just the portion of the price above the threshold. For example, on a £250,001 house, you pay the £250,001 rate on the entire purchase price.

LAND REGISTRY FEES

Fees are based on the cost of the purchase and are currently as follows:

Value	Fee
less than £50,000	£40
£50,0001–£80,000	£60
£80,001–100,000	£100
£100,001–£200,000	£150
£200,001–£500,000	£250
£500,001–£1m	£450
Over £1m	£750

Budget and finance checklist

AVAILABLE FUNDS

- [ ] Make sure you know exactly how much money you have readily available, and the absolute maximum you can stretch to.

- [ ] Decide whether you are selling your home in order to by the new property/land.

- [ ] Find out if you will be liable for capital gains tax.

- [ ] Write a detailed list of all of the selling/purchasing costs.

BUDGETARY CONSIDERATIONS

- [ ] Calculate the size of house/plot of land according to your available funds.

- [ ] Consider how the location of the land may affect its price.

- [ ] Choose a design/structure that suit the funds you have available.

- [ ] Allow for a contingency fund of at least 10–15 per cent.

POTENTIAL SAVINGS

- [ ] Decide on the quality you want for fixtures and fittings.

- [ ] Shop around for prices on land and location.

- [ ] Get competitive quotes (three) when it comes to tender.

- [ ] Find out whether you can claim back any VAT.

- [ ] Make sure you have a structural warranty and site insurance.

KEEPING ON TRACK

- [ ] Hire professionals (project manager, architect, quantity surveyor, building contractor, labourers) to smooth the way.

- [ ] Compile a cost plan to monitor each step of the build.

- [ ] Have all quotes and costs confirmed at an early stage.

- [ ] Keep track of your working costs.

- [ ] Re-allocate funds as the project progresses.

- [ ] Keep track of costs against the budget.

2 Finding the perfect spot

Every self-build project, from the simplest renovation to the most complex new build, starts with the process of finding a suitable property. For some, this is as easy as peering out of the window at a large side garden and realizing there's a plot on the doorstep; for others, it involves looking further afield.

Using the local resources

In areas of good supply, a visit to the local estate agents, or scouting the area you are interested in, looking for 'for sale' boards, is the best way to start looking for land.

ESTATE AGENTS

Estate agents sell most building plots, as well as properties for renovation, demolition and replacement or conversion. Looking for a plot is really no different from looking for a house to buy – you need to work out roughly what you want, register with a number of estate agents and keep in touch with them.

In any given area there will only be a few key agents who handle plot sales on a regular basis, and these are the agents to target. You will be competing with professional builders and developers, so visit their offices to let them know your requirements. Leave them something on paper, including your address and contact numbers. Let them know you have your finances organized, and that your solicitor is ready to act quickly when the right plot comes up.

Having made contact, stay in touch at regular intervals to make sure that you hear about any potential sites before they sell. Do not expect too much from the agents – you could be one of hundreds on their books – and, remember, agents work for sellers, not buyers.

Having enlisted the services of an agent, bear in mind that building plots come up in other places too. Sometimes they are advertised in local property papers, often by an estate agent. Each month self-build magazines feature hundreds of plots available, nationwide, in their land listing pages.

BUYING AT AUCTION

Many plots are sold at auction. Held nationally, regionally and locally, you can find out about them through local advertising, on 'for sale' boards or auctioneers' websites. In many cases they are organized by estate agents, in which case have your name added to a mailing list for auction catalogues.

COUNCILS AND LARGE COMPANIES

Once in a while, a local council will sell off building plots, and it is worth contacting their estates departments. The same goes for large companies, particularly utility companies (water, electricity, gas): they have land holdings, portions of which get sold off from time to time. Much of this land ends up being sold through estate agents or at auction, but you might have a chance to buy before this happens.

Enlisting the help of others

In areas where self-build opportunities are scarce, it is advisable to apply the search techniques that builders and developers use to locate, or even create, the ideal site. You may need to call in the professionals.

SALES JARGON
OPP Outline planning permission
DPP Detailed planning permission
OIRO Offers in the region of
OIEO Offers in excess of
POA Price on application (to the agent)

Looking for a plot of land can seem an altogether more daunting task than looking for a house, particularly as one plot may look very much like any other.

LAND-FINDING AGENCIES

These are companies that maintain databases of land for sale, and to which you can subscribe. They either send you details of suitable plots or you can search their websites on the Internet. For a modest fee these services provide a wealth of information about plots coming on to the market, who is selling them and what the asking prices are.

SELF-BUILD PACKAGE COMPANIES AND BUILDERS' MERCHANTS

Self-build package companies can provide help in various ways, from maintaining land lists to actively searching for suitable sites. Builders' merchants often have self-build clubs and circulate lists of plots to their members. A local builders' merchant is also a good place to put a 'land wanted' card.

EMPLOYING AN AGENT

A few estate agents specialize in dealing with building land and will look for plots on your behalf. This is done in return for a commission of 1 or 2 per cent, which you pay if you buy a property they introduce to you. Alternatively, you might see specific plot-finding services advertised. The golden rule with any plot-finding service is to be clear about precisely what you are paying for and when payment is due. Avoid any service that obliges you to pay, regardless of whether you buy land or not.

USING A PLOT HUNTER

While the term 'plot hunter' is not generally used in the property profession, it is a useful label for consultants who advise on land finding, planning permission and site suitability. Few professionals offer this range of services because it combines the expertise of an estate agent, a planning consultant and a building surveyor/architect. If you need someone to help you analyse your plot, try to find a planning consultant who is also a chartered surveyor and therefore has the required breadth of knowledge. As with other professionals, a recommendation is often the best way to find a suitable consultant.

NETWORKING

You can engage your friends, colleagues and family in your search, either just by letting them know you are looking or by offering them a reward if they find you a plot. A sum of £500 or even £1,000 is a small percentage of the average self-build budget, so well worth paying if it gets you what you want.

Finding land that is not on the market

There is nothing to stop you spotting and exploiting new opportunities for yourself. Builders and developers do this as a matter of routine, and you can use their methods. One option is to approach them direct. They often keep land banks and are, from time to time, prepared to sell off a single plot or a plot on a larger site they are developing. You might encourage them to part with some land by offering them an incentive, such as getting them involved in your build.

With each plot you see, try to visualize how you might use the land. Take note of buildings in close proximity and consider how they would influence any plans.

LAND-FINDING PROS AND CONS

	Pros	Cons
Estate agent	• Largest resource • Free	• Many do not sell land • No obligation to buyer • Time consuming
Auction	• Possible bargains • Unusual properties	• Research time limited • Problem properties • Up-front legal costs
Land-sourcing websites	• Time efficient • Deal with estate agents direct • Some land exclusive	• Charge for services • Not all agents participate • Information not always up to date
Councils/large companies	• Early notice of sales • Occasional prime sites	• Difficulty accessing information • Limited choice
Self-build companies	• Vested interest in helping	• Limited contacts • Most not set up to find land
Builders' merchants	• Useful forum for information	• No dedicated service
Self-sourcing	• Free • Spot bargains • Direct contact with owners	• Time consuming • Knowledge needed • No guarantees of sale

SEEKING PROFESSIONAL HELP

Professionals, such as architects, building surveyors and planning consultants, handle applications for new houses and so might be able to put you in touch with owners who are willing to sell. You could be lucky and do a deal direct with an owner; at the very least, you will be aware of an opportunity before it goes to an estate agent, putting you ahead of the market.

USING COUNCIL RECORDS

All new houses require planning permission and there is nothing to stop you looking through lists of planning applications and approvals at the local council offices or on its website. There will be detailed maps showing where applications have been made and planning permission granted. You can see, for example, whether applications have been made on infill or back garden plots in a particular area. Records will show whether permission was granted. These records are public information, and include a description of what has been applied for, together with contact details for the owner or their professional representatives.

Many applications for new houses are made by people wishing to build for themselves, while others will simply be hoping to get planning permission and sell on. A few phone calls is all it takes to find out the circumstances. Or why not visit the sites for which permission was granted: although most of the permitted houses are likely to have been built, there may be one that has not.

LOCAL PLANS

If you are looking at council records, take a look at their local plans at the same time. These are detailed documents comprising written policies and maps showing where the policies apply (see page 40). Policies cover everything from where new house building can take place, to the criteria applied to conversion and demolition, to replacement applications. Restrictive policies apply in areas such as green belts or areas of outstanding natural beauty (AONB). More locally, restrictive policies also apply in conservation areas and other local designations such as areas of high townscape value, and areas of special character. Use the maps to locate where the restrictive policies apply and to help target your search in areas where new housing is likely to be approved. Some councils also publish quite detailed supplementary planning guides on anything from extensions to rural building conversions. These give a useful insight into the council's thinking on more detailed matters of design.

LOOKING ON THE GROUND

You can take a more direct approach and go in search of land where a new house could be built or where there are buildings suitable for conversion. In towns and villages this might be a front, side or back garden, a small orchard or paddock. It could be parts of two or even more gardens. Also, look for derelict, unused sites or closed-down business premises.

Driving, cycling or walking around gives you part of the picture, but it often helps to consult Ordnance Survey maps to see exactly where there

Existing buildings on a plot can offer great opportunities for development. Before committing, however, investigate the local council's stance on your proposal.

are suitable spaces to build. Trees, hedges, walls, undergrowth and existing buildings can all disguise possible opportunities when you're just looking on the ground.

DEMOLITION AND REBUILD OPPORTUNITIES

In the countryside, opportunities to create new-built homes are likely to be limited to demolition and rebuild. Old cottages and farmhouses, wooden bungalows and chalets are all suitable candidates for replacement. Beware, however, of derelict houses. If the building is too far gone, the council might not let you replace it. In some areas councils restrict the size of any replacement to something not much bigger than the original.

POTENTIAL CONVERSIONS

Farm buildings, barns, coach houses, mills, and even pubs, might be suitable for conversion. You might also come across something like an old scrap yard, where there is potential to give up the existing use in exchange for a new house.

FINDING THE OWNERS

Once you locate a promising looking site or building, you then have to contact the owners to find out whether they are interested in selling. This can be as simple as knocking on a neighbouring door, or asking at the local pub or post office.

Failing that, council-planning records will help if there has ever been an application on the property.

Don't overlook plots of land that appear to be in use. You might find that a local farmer has a field or paddock that he or she is willing to sell.

If the property in question is a house, try the electoral role. You can also get details of properties from the Land Registry, for a small fee, although not all land is registered.

Site potential

Finding the right plot, or building to convert, is obviously a vital step in every self-build project. You might have located a likely looking opportunity, but how do you know whether it is really suitable, or that there are no pitfalls to upset the viability of your project? At this stage you need to look for anything that could prevent you building the home you want – physically, legally or financially – and you need to understand the potential problems to avoid getting caught out.

SIZE OF PLOT

There is no maximum or minimum size of plot, or any rule of thumb for calculating a required area. A plot only has to be big enough to accommodate the width and depth of the house you want to build (assuming the house is compatible with neighbouring houses), plus, usually, a private garden area, car-parking spaces and room for maintenance access around the building. Some councils have detailed policies prescribing minimum garden space and space between houses, and sometimes these details are contained in supplementary planning policy guidance. While it is common for most self-builders to begin by looking for a site that is around 1,000 m² (0.25 of an acre), they usually have to end up modifying their requirements in accordance with the land available.

SITE BOUNDARIES

Get on to the site itself and walk over it thoroughly. First, look at whether the house you want will fit in. If space looks tight, get a tape measure and check. If the boundaries are not clear, you will have to check their position from the sale plans. Do the boundaries on the ground tie up with the boundaries on the sale plan? Do these tie in with the area that has planning permission? Do these match up with the deeds? Where discrepancies come to light, call a meeting on site with the vendor, estate agent and both parties' solicitors to thrash it out. Do not sign a contract until you are certain of what you are getting.

POSSIBLE OBSTACLES

Look for any obstacles that could stop you building where you want. Ponds, ditches, trees, outcrops of rock, existing buildings, areas of concrete are all fairly obvious. But also look for rough ground or rubbish that could indicate tipping or made up ground. Reeds and rushes, willows and alders can all indicate wet areas or springs that might be dry in summer. Services, too, can be over or under the land: manholes, meters, stopcocks all need investigating. You can divert services, but only at a cost, so check their exact position both on the ground and with the relevant service company.

PATHS, PIPES AND CABLES

Watch out for any tracks across the site or stiles or gates on the boundaries. Public footpaths and bridleways are usually well marked, but private rights of way are not. If you see anything that looks like a footpath, make a note to get your solicitor to check it out. Similarly, manholes, or cables over the plot, signal the likelihood of an easement, or right of entry, held by a service company, to get access to and maintain their equipment. Again, get your solicitor to investigate the situation for you. There is often an exclusion zone around pipes and cables, which can restrict where you can build.

Make sure you are aware of anything on your plot that might affect your project – ambiguous boundaries; poor access; the positioning of any trees.

TREES ON THE SITE

Look carefully at any trees on the site and jot down their positions, heights, spreads and species. Are they in the way? Will they block light to the house or garden? How close will they be to the house? If you think a tree will need to be felled check with the council first, to see whether it has a tree preservation order (TPO) (see page 52). Trees, and their removal, can affect foundation design and, therefore, have build-cost implications.

SITE ACCESS

You must check whether there is adequate access to the plot. Have a look at the roadside boundary. Is there direct access on to a public highway? Be aware of private roads – speed bumps and poor maintenance are telltale signs – and for intervening land, such as a wide verge, between the plot and the kerb. Land in another ownership, whether private or council, means you need a right of access over it, and that can be expensive. Get your solicitor to confirm who owns what.

Also look to see if there is good visibility at the point at which your drive would join the kerb. Judge this by taking two-and-a-half paces back from the kerb and looking up and down the road. How far can you see and is there anything in the way? Hedges, fences, trees and telegraph poles are likely offenders, as are bends in the road or the crest of a hill. Councils have standards for visibility, which vary depending on traffic speeds in the road (see below).

SAFE VISIBILITY DISTANCES AT ACCESS

From the centre of the access drive, adjoining the kerb, take two-and-a-half paces back into the site. The distances of clear visibility up and down the road should be:

Speed limit	Distance
48 kph (30 mph) – estate road	60 m
48 kph (30 mph)	90 m
64 kph (40 mph)	120 m
80 kph (50 mph)	160 m
97 kph (60 mph)	215 m
113 kph (70 mph)	295 m

If visibility along the road is poor, ascertain whether it could be improved. If cutting back the neighbour's hedge is the answer, you cannot assume he or she would be happy to do this free of charge, if at all.

SERVICES ON THE SITE

While standing at the roadside, check for services – water, gas, electricity, telephone or sewers – in the road or verge. You will need to connect up to these eventually, so note where they are, bearing in mind you have no automatic right to cross someone else's land to get to them. Sewers, in particular, can be inaccessible, and although private systems, like septic tanks, might solve the problem, these can have implications for the site layout.

THE EFFECT OF YOUR SURROUNDINGS

Now is the time to take a closer look at the surroundings of the plot, with an image of your built house in mind. Do neighbouring properties overlook the garden, or vice versa? Will the garden area be shaded or sunny? Where are neighbours' drives and parking areas? Any of these factors could alter where you want to build. Take account, too, of neighbouring uses. Is there anything noisy or smelly in the vicinity that could affect your enjoyment of the plot? Noise and smell levels can vary greatly with weather and wind direction so check at a variety of times and conditions.

With your ideal house position now in mind, make sure other elements, such as your garage, drive and turning area will all fit in. These tend to push a house back into its plot – would this matter in your case, or could this interfere with trees, services, or a neighbour's privacy, for example?

MAKING GROUND SURVEYS

Further to your on-site checks, you must establish ground conditions by means of a soil survey. Although there are foundation solutions to overcome most poor soil conditions (rafts and piles), these can be costly and could upset your budget. Worst of all, they can upset it at an early stage in the project.

Poor soil conditions include heavy clay, running sand, and made-up or tipped land. You might recognize these yourself if you dig a few experimental holes about the plot (with the owner's permission). Check also with a building inspector from the local council's Building Control Department. These people spend much of their lives peering into trenches excavated for foundations and might well be able to give an accurate indication of likely soil conditions on the plot. A soil survey can be anything from a soil scientist taking samples for analysis in a laboratory (and possibly recommending a necessary foundation type) through to getting in an excavator to dig some holes and having your architect, builder or anyone else with a knowledge of soils and building to have a look. The former has the advantage of accuracy, the latter, low cost.

FLOOD-RISK AREAS

A final, but very important, check on site relates to flooding. You can find out whether the plot is in a flood-risk area from the Environment Agency and by looking at the council's local plan. On site, though, be alert for any localized risk, such as low-lying ground, or the close proximity of a large slope where water run-off could be a potential

PLOT ASSESSMENT CHECKLIST

These are the things you need to consider when assessing a plot of land.

On site	Off site
Size	Deeds – boundaries
Boundaries	Planning permission
Obstacles	Planning file
Paths/rights of way	Covenants
Pipes and cables	Rights of access
Trees	
Access – adequacy of	
Services – availability	
Neighbours	
Surroundings	
Soil survey	
Flood risk	
Site survey	

Flooding damages and destroys property. It is essential that you check whether your potential plot is located in a flood-risk area.

hazard. You should talk to several local people to ensure you detect any site-specific risk.

THE VALUE OF A SITE SURVEY

Where a plot is an odd shape, sloping, or has a lot of trees on it, it might not be obvious where or how your new house will fit in. In these cases, you should commission a site survey to provide an accurate and detailed site plan. Some architects and building designers carry this out as part of the service.

Land surveyors can produce an accurately surveyed site plan, including details of levels, for a cost of about £500 to £1,000, depending on the size and complexity of the plot. This might sound expensive, but an accurate survey is invaluable for pinning down exactly what you're buying, for designing the site layout, in negotiating with planning officers and, ultimately, when you come to set out the foundations.

WHEN TO BUY AND WHEN TO WALK AWAY

It is not always possible to get a definitive answer to every question about a property before you buy. It is essential, therefore, to sift through all the issues surrounding your purchase and identify any unknown factors that could potentially add significant cost to your build or, worse still, prevent it all together. Use the checklist on page 34 to ensure you have got all the possibilities covered.

Much depends on the state of the market. In a slow market, where supply exceeds demand, you can afford to take your time. You can check everything thoroughly, take professional advice where necessary, and possibly negotiate the price down if you find unexpected costs. In a busy market with plenty of buyers and rising prices, however, you have got to be much more careful.

Do not let your desperation to secure a plot cloud your judgement. If there is an outstanding issue that could jeopardize the whole project – suspected right of access or planning difficulties – make your offer subject to resolution of that issue. If the vendor will not accept such a condition on your offer, walk away. The golden rule is not to be in a position where you are obliged to pay for something that should actually be the responsibility of the vendor. Poor ground conditions, restrictive

covenants, planning constraints or drainage problems all affect the value of the land. If a vendor tries to pass these costs or problems on to you, do not buy.

Planning and legal checks

In addition to your on-site checks, you must look through the deeds and planning permission for anything that could prevent or add cost to your build. This is a job for your solicitor.

Restrictive covenants and conditions attached to the planning permission might restrict where a house can be built, its size or design, or even whether a house can be built there at all. Where these affect the site layout, refer back to your on-site assessment to ascertain whether any change might have wider implications.

PLANNING CONDITIONS AND LEGAL AGREEMENTS

Planning permission for a new house is always granted subject to a number of conditions, for example, approval of drainage details, landscaping schemes, specific trees to be kept or protected during the build, and approval of access. Occasionally, planning permission is also accompanied by a legal agreement, known as a 'Planning Obligation' or 'Section 106 Agreement'. This might, for example, restrict who can occupy the property (such as someone working in agriculture only) or require a payment to the council as a contribution to local schools, libraries or transport schemes.

COVENANTS AND RIGHTS OF ACCESS

Problems with covenants and rights of access can usually be resolved by negotiation with whoever owns the access or has the benefit of the covenant. Substantial sums are paid to release restrictive covenants or gain access, and these costs affect the value of the land. Consequently, you must try to uncover them as part of your site assessment so they can be accounted for in your offer for the site.

In some cases the owner of a private road cannot be found, or the property that had the benefit of an old covenant is long demolished. It is possible to get indemnity insurance in such cases, where you pay a one-off premium of several hundred pounds. Your solicitor or an insurance broker can arrange quotes.

CHOOSING A PROFESSIONAL

Make sure you choose the right professional for the right job.

Profession	Areas of expertise
Chartered planning and development surveyor	Land assessment and planning
Chartered town planner	Planning
Architect	Design, building regulations, build management
Chartered building surveyor	Design, building regulations, build management
Architectural technician	Plan drawing

FINDING A PROFESSIONAL

There are many different ways to find a professional. The following list is in order of preference.
- Personal recommendation.
- Similar jobs locally (this information can be sourced from planning records).
- Professional body: Royal Institution of Chartered Surveyors (RICS); Royal Town Planning Institute; Royal Institute of British Architects (RIBA).
- Business telephone directories/local advertisements/Internet search.

QUESTIONS TO ASK PROFESSIONALS
- What experience do you have?
- Can you show me examples of similar cases?
- What is the exact scope of the work?
- What timetable do you expect to work to?
- What kind of fee basis do you work to, and when are payments due?
- Can you give me names of local jobs/clients for references?

Maximizing plot potential

When looking at a potential plot, try to find opportunities for making more of it than perhaps the current planning permission allows. If there is permission for one house, could the plot take two or more? If a bungalow is permitted, could you get additional rooms in the roof or even build a full two-storey house? Could a three-bedroom house be extended to four or five bedrooms?

Bear in mind that a plot is likely to be priced on the basis of either a detailed planning permission or an outline (see page 39). If the plot has detailed permission, the price will probably be based on just the permitted scheme. If the plot has outline permission, the price will be based on an assumption about what can be built.

Any improvement you make is going to add value. This may mean you can bid higher than others to secure the plot if your budget allows. Alternatively it means that you buy at a bargain price. Maximizing potential can be a matter of innovative design, effective problem solving or building good relationships with neighbours, planners and local councillors.

Assessing the potential for a conversion

All the considerations that apply to plot assessment apply equally to conversions except, of course, the building position is largely fixed. However, there might be scope to extend or build a garage or annexe.

SUITABILITY OF THE BUILDING

A vital extra assessment job for conversions and renovation projects is to check the size and condition of the building to ensure it is suitable for your purpose. You will probably want to get expert help with the finer details of this, which means employing a building surveyor, or possibly an architect, to undertake a structural survey.

WAYS OF MAXIMIZING POTENTIAL

Potential	Buildings	Land
Extend back, front and sides	✓	
Larger house		✓
Sub-divide into smaller units	✓	✓
More units on plot		✓
Demolish existing buildings to create more build space	✓	✓
Less parking to create more build space		✓
Improve internal layout	✓	✓
Add room in roof	✓	✓
Add room in basement	✓	✓
Add whole new floor	✓	✓
Change use	✓	

Some councils only allow the conversion of buildings with historic merit, where conversion is a means to preserve them. Find out from the council whether the building is listed and, if so, what special features are recorded, as you'll undoubtedly have to preserve them (see page 51). To prove the historic worth of a building that isn't listed you might need to employ an archaeologist to undertake a survey and research the building's history. This would apply particularly where the council does not consider the building sufficiently interesting to be worth preserving by conversion.

PLANNING POLICY RESTRICTIONS
While there is endless potential for the conversion of buildings in towns and villages, out in the countryside planning policies are making life difficult. Essentially, the Government now prefers rural buildings to be reused for business, tourism or community purposes, rather than be converted to houses. It is usually necessary to show why a building cannot be used for any of these alternative uses before the council will consider conversion into a house. This might involve marketing the building for six or nine months for business use. You can sometimes avoid this if you can show, for example, that the access is unsuitable for business traffic, or that the building has other houses very nearby.

PART BUSINESS, PART RESIDENTIAL SCHEMES
Some councils have policies that allow residential use where it is a 'subservient' part of a business scheme. This means the floor area of the business use must be over 50 per cent of the total floor area of the building. If you have a group of buildings, the 50/50 approach can be a particularly useful way of overcoming the current resistance to rural conversions.

MAXIMIZING THE POTENTIAL OF A CONVERSION

Although councils often resist extending rural buildings as part of a conversion scheme, extensions can sometimes be achieved. Research into the history of a building might reveal it was larger in the past, and your extension can be presented as an opportunity to restore the building's former glory. A sympathetic design and use of appropriate materials are essential. In addition, you can sometimes trade off the removal of other, perhaps modern or unattractive buildings, for a bigger extension to your conversion.

Assessing the potential of a renovation

Renovations can range from a new coat of paint to a virtual rebuild. If a building is run down you will need to get it surveyed to ensure it is structurally sound and free from damp and dry rot. Unless the building is listed, you have free rein to make internal alterations without the need for planning permission.

External changes will be tightly controlled in conservation areas or if the building is listed. Otherwise, most minor external changes will probably be permitted development (see page 49) and will therefore not need planning permission. More significant external changes, such as a large extension or roof alterations, are likely to need both planning permission and Building Regulations Consent and you need to factor into your budget the time it takes to obtain these.

Assessing the potential for demolition and rebuild

Most buildings can be demolished to make way for a rebuild. The exception to this is a listed building, where you are unlikely to get planning permission to demolish it.

In conservation areas, Conservation Area Consent for demolition is needed. Again, if the building in question makes an important contribution to the character of the conservation area, permission is unlikely to be forthcoming. Similarly, if a building is considered by the council to be of historic or architectural interest, it might prove impossible to get planning permission for demolition and rebuild.

Ambitious conversion and extension schemes can be achieved with a little knowledge of the original building's history and a well-considered proposal.

REBUILD RESTRICTIONS

Even if you find a house that you can demolish, it does not necessarily follow that you are allowed to rebuild it. Difficulties can arise where a rural house or cottage is derelict: if the condition of the house has deteriorated too far, the council might decide that its residential use has been abandoned and so cannot be replaced. There is no precise definition of abandonment, but the courts have a set a number of tests. These include the length of time the building has been vacant; its structural condition; whether there has been any intervening use – such as a farmhouse being used for storage or as a cow barn; and the intentions of the owners. Even within these guidelines interpretation varies.

A reliable common-sense test is to ask yourself whether the building is still recognizably a house or is simply a ruin. If the latter, the chances are the council will consider the residential use abandoned and not allow a replacement. If the former, then, assuming there has been no alternative use of the building, you have a good chance of getting permission to replace it.

DEMOLITION COSTS

Once you are satisfied that the house can be demolished and replaced you should carry out the same site checks as for a plot (see pages 32–35), with the added consideration of demolition costs. At best you will have valuable salvage materials to sell at a profit, or possibly materials for recycling into your new build – either into the house itself or in the form of hardcore under any new drive or parking areas. At worst you might have dangerous materials, such as asbestos, that require specialist contractors to come in and remove them.

MAXIMIZING DEMOLITION AND REBUILD POTENTIAL

Council policies vary but most expect a rural replacement house not to be significantly larger than the original and to be situated in more or less the same place within the site. The theory behind this is that the impact of the new house on the character of the countryside should be no greater than that of the original. With this in mind, it is possible to argue that a larger house with a lower roofline would be less obtrusive, or that the addition of a basement to a replacement house would make no difference to its outward appearance.

If an existing house is built close to a road, moving the new house further away might help justify a few extra square metres. As with conversions, removal of other buildings on the site can be a useful bargaining point to gain extra room in the building. A pair of old cottages might justify one larger, single house. The justification here is not just confined to the size of the building but also to the amount of activity generated. One house would produce less traffic than a pair of cottages, and this might be viewed as a plus point that would justify an increase in the size of the replacement.

Planning permission

The majority of self-builders either have to get full planning permission because they have bought a plot with outline permission only, or want to change an existing permission, or wish to get permission where none exists at present.

WHAT IS PLANNING PERMISSION?

Planning permission is the authority required by law to carry out development. This includes new building work and changes of use, such as converting buildings into houses. Planning permission alone, however, does not mean you can build. You also need Building Regulations Consent and the land in question must be free from any legal restraints, such as covenants or restrictions on access. There are other consents that might be needed in specific

circumstances: any work affecting a listed building (see page 51) must have Listed Building Consent and demolition in a conservation area needs Conservation Area Consent.

WHO GIVES PLANNING PERMISSION?

The planning system is run by a combination of central and local government. Day-to-day planning issues are dealt with by local councils. For simplicity, this book uses the term 'district council' to cover all local planning authorities. In many areas there are also parish, community or town councils, which, for convenience, we refer to as 'parish councils'. These councils do get consulted on applications but do not have any formal role in administering the system.

HOW ARE DECISIONS MADE?

The law prescribes that decisions on planning applications must be made in accordance with the district council's 'development plan', unless so-called 'material considerations' indicate otherwise. The following section looks at development plans and then considers the various aspects that do, and do not, qualify as 'material considerations' that can influence a decision.

DEVELOPMENT PLANS

The development plan system is broadly similar across the UK and the Republic of Ireland. Councils draw up plans containing planning policies, which cover issues like housing, employment, shopping, recreation and environmental protection. Structure plans are drawn up by district councils and contain broad strategies for the whole county, while local plans are much more detailed, comprising a written document of policies with explanatory text and maps, to show where the various policies apply. The individual authorities combine both structure and local plans together into a single 'unitary development plan'.

LOCAL PLANNING POLICIES

Local plans vary from council to council but all contain detailed polices on the quantity and location of new housing needed in their district. Maps show the boundaries between towns and villages and countryside. These are given a variety of names such as 'settlement boundary', 'built-up area boundary',

SAMPLE PLANNING POLICIES

Below are some example of planning policies. Remember that policies will vary depending on the council.

Type of housing/location	Criteria to be met
Urban housing: new estate development	• To protect the established character of the existing residential area, there must be a mix of dwelling types, with a minimum density of 30 dwellings per hectare.
Conversion and extension of urban housing	• The style, design and materials must be appropriate and sympathetic in relation to the existing buildings. • There can be no significant adverse effect on the amenities of the occupiers of neighbouring properties.
Rural village housing	• Development cannot lead to a stark or less loose-knit definition between the village and its surrounding countryside.
Conversion of rural buildings for residential use	• The applicant must have made every reasonable attempt (and failed) to secure suitable business reuse. • Residential conversion is a subordinate part of a scheme for business reuse.
Rural demolition and replacement	• The proposal must be of comparable size and massing to the existing building. • It must be in keeping with the character of the locality. • It must be similarly sited within the plot, unless an alternative siting would result in landscape, highway safety, or local amenity benefits. • There can be no loss of a building of local character.

'village envelope', 'housing framework' and 'built confines'. These different terms are used to refer to the same thing, namely lines marking the change from areas where restrictive policies to protect the countryside apply to built up areas where additional building is more likely to be allowed.

There are usually specific policies on infilling and small-scale development in built-up areas, and on extensions to houses, demolition and replacement, and conversions in the countryside. Policies also cover building in designated areas such as green belts and AONB and, in towns and villages, conservation areas.

Increasingly you can access council plans on the Internet. Because the process of amending and updating plans is very slow, there are often two plans available at any one time. The current, adopted, plan is the one that should carry the most weight, although councils often start using the policies of a new, draft, plan before it has been through the formal review process. If in doubt about council policies, ask a planning officer which policies are being used for day-to-day decisions.

Many councils also produce supplementary planning guidance, fleshing out areas of policy in the local plan. These documents, often published as booklets, do not carry the same weight as the formal policies when it comes to decision-making but they do represent an important 'material consideration'.

Select an appropriate building style: planners may object to a proposal if they feel it is not in keeping with the local environment.

MATERIAL CONSIDERATIONS

In some circumstances, the normal ruling of the local authority is overridden and planning permission is granted where, in the majority of cases, it would be refused. These circumstances are referred to as 'material considerations'.

GOVERNMENT PLANNING POLICY

National planning policies are produced by the Government and introduce influential 'material considerations' to weigh alongside local policies. Because of the slow local plan review process, local policies often lag behind those of the Government. In the UK, for example, the current drive by Government to increase the density of new housing is not always reflected in local council policy. Generally, national planning policies on housing and on development in the countryside are of most relevance to self-builders.

PLANNING HISTORY

A building's planning history can influence later decisions on an application: it might have an existing use, previous planning permission, or a previous refusal of planning permission, for example. Where there is already permission, you might be able to improve on it by producing a better design, or by overcoming a certain problem. Alternatively, a string of refusals on a site sets a negative precedent, which could be difficult to overcome. Previous planning appeals always have considerable influence, so get a copy of, and read, any appeal decision letter in order to find out what the issues were.

EXISTING USE

The type of buildings and/or the uses currently on a site are significant factors by which any conversion or new building will be judged. The classic example is that of an existing scrap yard in an attractive rural area. Such a use might be noisy, unsightly and generate a lot of heavy traffic on narrow country roads. An application to close it down and replace it with a well-designed house might well be approved, regardless of the fact that the idea of building new houses in the countryside runs directly against policy. Here, the existing use is such a compelling 'material consideration' that it can override development plan policy.

PHYSICAL FACTORS INFLUENCING DECISIONS

There are various physical factors that can influence planning decisions. To some extent these are the same factors that you would assess when checking a plot to see whether it is suitable for your particular project (see pages 32–35).

Architectural salvage yards are a great source of fixtures and fittings should you be looking to renovate an older property in keeping with its original style.

FACTORS INFLUENCING PLANNING DECISIONS

In planning terms:
- development plan policy
- special designations: conservation area; area of outstanding natural beauty
- supplementary planning guidance
- government planning policy
- planning history of site.

In physical terms:
- size and shape of site
- topography
- ground conditions: liability to flooding, subsidence, contamination
- archaeological remains
- boundaries: removal/preservation of boundary features
- availability of services
- trees and hedges: loss of, distances from, importance of
- wildlife: effect on rare or protected species
- access: meeting safety standards, visibility, adequate width
- access: appearance of, effect on trees/hedges, disturbance from
- car parking: provision of and on-site turning space
- rights of way: obstruction of, scope of relocation
- neighbours: preservation of privacy, avoiding overlooking, avoiding loss of light, avoiding overbearing or obstructive development
- surrounding area: character of area, sizes of buildings, position in plot, plot sizes, relationships between buildings.

LOCAL POLITICS

Whenever planning applications are made, neighbours, parish councils and district councillors are consulted. There is no doubt that the level of local support or objection can have a significant bearing on the outcome of an application.

The most popular reasons for neighbours to object to applications for new houses are the following: loss of a view, noise during construction, boundary disputes, motives of the applicant and affect on value. (None of these is actually relevant to a planning decision.)

OUTLINE PLANNING PERMISSION

Many building plots are sold with outline planning permission. This tells you whether, in principle, a new house can be built. As a general rule you cannot make outline applications for change of use or conversions. Also, councils do not usually accept outline applications in conservation areas. In outline applications, details such as the means of access, position of the house on the plot, size and shape of the building, and floor layout are usually omitted from the scheme, although one or more are sometimes included. The remaining, omitted, details are known as Reserved Matters (see below).

When should I apply for outline permission?

An outline application is usually made when the prospects for getting permission are uncertain or when the owner wants to sell the site in question. Where the outcome is uncertain, an outline application saves all the costs of getting detailed drawings prepared. Furthermore, if you want to sell, having an outline application will not restrict purchasers thinking about what could go on the plot. An outline permission leaves them free to devise and apply for their own scheme. Making an outline application also has advantages when the size or design of the house you want is likely to be controversial (see 'Reserved Matters', page below).

What are the fees for an outline application?

A council's outline application fees are based on the area of land involved. A standard fee (around £220) is levied per 0.1 hectare (about ¼ acre). On a really large site, significant application fees could wipe out the saving in design fees. However, on a large site, you could reduce the fee by defining part of the site as garden and part, say, as paddock. The fee would only be levied on the garden area but you could not turn the paddock into garden later without obtaining planning permission.

RESERVED MATTERS

Any details of a scheme not established under the outline permission can be submitted to the council as a Reserved Matters application. The details must not conflict with the description of the development on the outline permission, or with any conditions imposed at the outline stage. For example, you could not apply for a three-storey house where the outline permission was for a two-storey house.

Approval of Reserved Matters, together with an outline permission, is the equivalent of a full permission. The advantages of applying for approval of Reserved Matters is that the principle of building is not in question and you can make as many Reserved Matters applications as you like. The cost is a standard sum, based on the number of houses concerned (around £220 per unit).

FULL PLANNING PERMISSION

Full or detailed planning permission is, as the name suggests, planning permission for all the details of a specific house and plot layout, extension or conversion. You would normally make a full application for all types of extension or conversion and for demolition and rebuild projects. For a new house, a full application would be appropriate where there is reasonable certainty of permission being granted and the principle of building is not in question. Councils usually insist on full applications in conservation areas. There is a standard fee of around £220 per house or flat, regardless of its size or the size of the garden.

BUYING WITH PLANNING PERMISSION

If you are buying a plot that already has planning permission, make sure you obtain the planning permission documents as soon as you can. You must ascertain that the permission is still valid and will let you build what you want or will allow you to make changes. The estate agent or vendor should be able to supply photocopies of the permission documents (application forms, plans and drawings and the council's decision).

CHECKING THE PLANNING DOCUMENTS

The first point to clarify is whether the plot has got outline or full planning permission.
• Full permission is for a specific design and layout. Detailed drawings will have been submitted, showing site layout, floor plans and elevations.
• Outline permission establishes that you can build a house, but leaves the exact design and layout to be approved at a later date.

Some mortgage lenders insist that a plot has full permission before lending money on it. Planning permission applies to the site, not to the landowner or the person who made the application. This means you can make use of someone else's permission but be aware that you cannot build someone else's design without the designer's consent. If your site has full planning permission, but you don't want to build what is permitted, there is nothing to stop you putting in your own application for another design.

TIME LIMITS ON PLANNING PERMISSION

Planning permission has a limited life so check any dates on the permission carefully. With full planning permission, you must start work within five years from the date permission was given (the Government is talking about reducing this to three). With outline permission you have three years to submit details of the design of the house and layout of the site, plus a further two years to start work once those details have been approved. If the time limit has passed and no work has started, or details submitted, the permission has probably lapsed. In most cases you could get a renewal but there is no guarantee of this. If the seller claims to have started work in order to keep the permission alive get written confirmation of this from the local authority's Planning Department. Significant work must have taken place to be considered a genuine start of the work – the foundations being dug, for example.

CONSULTING COUNCIL'S FILE

If you are not given all the permission documents by the vendor or agent, it is worth having a look at the council's file, which you can read at their offices. There might be letters from other authorities, such as those governing highways and drainage, which give useful information. Look at the planning officer's report on the application. This can also give an insight into how the council would like to see the site developed and what factors you will have to take into account in your layout or design, such as preserving neighbours' privacy, position of access, or location of the house on the plot. An increasing number of councils are making this information available on their websites, but this does not always give the full picture. Handwritten notes, or scribbles on the cover of the file itself, can be revealing.

KNOW YOUR BOUNDARIES

All planning applications must include a location plan. This is vitally important because the red line drawn around the plot on this plan defines the area to which planning permission applies. You need to be certain that the land you are buying includes the whole planning application site. Match the location plan against the legal title plan and against the boundaries you find on the ground. Where a boundary is not marked – for example, because the site is part of a garden and there is no dividing fence or hedge – agree the position with the vendor and have pegs banged in to establish the line.

CONDITIONS WITH PLANNING PERMISSION

All permissions have conditions attached to them and these are set out and numbered on the decision notice itself. Standard conditions cover things like the time limit for starting work and getting external building materials approved. Others might limit the building to a single storey; take away the rights you would otherwise have to extend the building later on; require an access to be created to a specified design; or, in the countryside, even restrict occupation to agricultural workers.

Make sure that you can meet all conditions, especially if they involve work outside the site – cutting back a hedge to improve visibility for access, for example. Sometimes permissions have other notes added to the decision – often called 'informatives' – saying, for example, that the council would not be likely to approve any extension proposals. These are not conditions but do give useful pointers for you to take into account regarding the potential changes you can, or cannot make, to the building site.

SEEKING ADVICE ON PLANNING

If you have any doubts about your planning permission, speak to a planning officer, preferably the one who dealt with the application. In fact, it is a good idea to discuss the site with the officer anyway. Tell the officer what you want to do with the site and make sure he or she confirms that the permission accommodates your scheme. Get crucial points confirmed in writing to protect yourself. If you do not like the answers, or the officer will not speak to you, take advice from an independent planning consultant.

Making a planning application

Planning application procedure is the same whether you make a full, outline or Reserved Matters application. The first step is to decide how much help you want with your application. Do you need help with detailed plans? Are the planning issues too complex for you? Are you simply too busy to devote the necessary time? If the answer to any of these questions is 'yes' you should enlist the services of a professional.

Your scheme must comply with development plan policies or, if it does not, you need convincing reasons why permission should be granted. In the latter case, you might need professional advice to make what is often a finely balanced judgement. Whoever is preparing your plans should take full account of all the physical aspects of the site and its surroundings, and should guide you as to what is, or is not, likely to get permission.

MEETING THE PLANNING OFFICER

Once you have pinned down what you want to build and who is going to help you, it is worthwhile meeting a planning officer. Unfortunately, not all councils are prepared to give pre-application advice and some will not even reply to pre-application letters. However, most have a duty planning officer available by appointment. Take a sketch scheme of what you want to build and ask if the officer foresees any insurmountable problems. This is a more positive approach than simply asking what the officer would expect or like to see built on your site.

It is possible that the planning officer you meet will not be the one who deals with your application when it is submitted. A different officer could hold very different opinions. Also, the officer probably does not have a detailed knowledge of the site and certainly is not in possession of all the facts for making a formal decision. Instead, use the meeting to flag up likely areas of contention or disagreement.

LOCAL LOBBYING

When you make your application, your neighbours are automatically notified by the council and asked for their opinion. As a matter of courtesy, it is advisable to let

APPLICATION FORM REQUIREMENTS

Councils produce detailed instructions with their application forms and you should read these carefully.

Application requirements	Full planning permission	Outline permission	Conversion/ extension	New build/ change of site layout
Location plan (Ordnance Survey extract) showing the extent of the application site edged in red and any other land in your ownership edged blue.	✓	✓	✓	✓
Detailed plans, showing all the elevations and the floor plans of the building.	✓		✓	
'As is' plans to show how the building is changing			✓	
Layout plan showing the means of access, parking and turning areas, and existing and proposed landscape features	✓	✓	✓	✓

them know of your plans before they hear it from the council. Ideally, engage them in the process, showing them the plans and ensuring that you have their support before you submit. Predicting and preventing objections to your application is an important part of the process. You could also approach the parish council and enlist its help in advance.

Filling in the application forms is usually straightforward but resist the urge to include too much information. Under 'proposed development' writing 'erection of dwelling' is quite adequate. Either attached to the application forms, or on a separate sheet, you will find various certificates of ownership. You must sign a certificate stating that you own the entire application site or, if you do not, that you have served notice of your application on the owner. Make sure you sign and date all the forms and supply the number the council requires – normally five – and remember the cheque for the application fee.

Several days after submitting your application, it is registered as being complete and given a reference number. The council writes to you with these details and sometimes gives the name and number of the officer allocated to the case. Pass these details on to anyone who is supporting your scheme, as now is a good time for them to write in. Following registration, the council sends details of your application to other interested parties, including the Highways Authority, the Environment Agency, the parish council and your neighbours. They are given approximately three weeks to respond. You should contact the council's planning officer towards the end of this period, partly to establish a rapport and partly to see if he or she has any comments.

Councils and officers vary greatly in their willingness to discuss applications. Key questions are whether your application is likely to be approved or refused, whether it will be decided by the officers or by a planning committee and the likely timetable. The circumstances in which an application is determined by committee vary from council to council. Typically it occurs where the planning officer

PLANNING APPLICATION PROCEDURE (ENGLAND, SCOTLAND, WALES AND NORTHERN IRELAND)

Timing	Action
Start	Application submitted.
Week 1	Application registered and given formal reference number. Details sent to interested parties.
Week 4	Results of consultations back to planning officer.
Week 5/6	Planning officer visits site.
Week 6	Planning officer writes report to committee.
Week 8	Decision issued by planning officer, or application goes before planning committee, who make decision (for Northern Ireland, see page 48).

PLANNING APPLICATION PROCEDURE (REPUBLIC OF IRELAND)

Timing	Action
Start	Notice published in local newspaper and site notice put up.
Week 2	Latest date for lodging application.
Weeks 2/5	Application validated, submissions and objections considered.
Weeks 5/8	Request for further details or notice of decision.
4 weeks from notice of decision	If no appeal made, permission granted.

is recommending approval but there are a number of objectors; where the local councillor requests a committee hearing; where the issues are finely balanced; or where the development is particularly contentious. There are no planning committees in Northern Ireland.

If the case is going to committee, you can find out in advance whether the officer is recommending approval or refusal. Five days before the committee meets, you can see the officer's report, which summarizes the case. It is worth obtaining a copy of this, as there may be inaccuracies you want to correct.

PLANNING COMMITTEES

Most councils allow applicants and objectors to speak at planning committee meetings, each having two or three minutes in which to make his or her case. Write down what you have to say, to avoid missing key points. You may also wish to list key points, or anything technical, beforehand and circulate it to the committee members. You can get their contact details from the council or its website. Applications are either approved or refused. Sometimes the decision is deferred, pending further negotiation, amendment or the completion of a legal agreement. Even if approved, check the conditions carefully for anything onerous.

If your application is refused, consider whether the reasons given for the refusal are reasonable and whether they can be overcome. Discuss the refusal with the planning officer. Are the reasons matters of detail that can be changed or are they a matter of principle? If there is scope to negotiate and resubmit your application, you can do so free of charge during the 12 months following refusal. If there is no other way forward, you can lodge an appeal within three months of refusal. Similarly, if permission is granted subject to unacceptable conditions, you can appeal against some or all of those conditions. The appeal system is somewhat different in the Republic of Ireland (see page 48).

PLANNING APPEALS
What are my options?

You can elect how you want your appeal to be determined, and there are no statutory fees. There are three options:

1 **Written representations** This is suitable for most appeals. It is the quickest option, with no possibility of costs awards. The appeal is determined by the exchange of written cases, followed by a site visit. There is no discussion on site.
2 **Informal hearing** This is suitable for appeals where discussion on site is important. The appeal takes two to three months longer than written representations, and costs can be awarded. The exchange of written cases is followed by a semi-formal discussion with the planning inspector. The case can be discussed at the site visit.
3 **Public inquiry** This is suitable where complex planning issues or points of law require debate. This will take two to three months longer than an informal hearing, and costs can be awarded. The exchange of written material is followed by a formal, quasi-judicial hearing. Barristers may be involved and there is an opportunity for cross-examination. Cases cannot be discussed during a site visit.

Who determines planning appeals?

Appeals are made to the Planning Inspectorate in England and Wales, the Inquiry Reporters Unit in Scotland, the Planning Appeals Commission in Northern Ireland and to An Bord Pleanala in the Republic of Ireland. Whenever planning permission is refused, you are automatically sent details of how, where and when to appeal. You can also access information, and in some cases, appeal forms, via these various bodies' websites.

How should I go about making an appeal?

Although you can fight an appeal yourself, you are up against the council's professional planning officers, who are likely to know infinitely more about planning law and practice than you. Take professional advice from an independent planning consultant.

First check the prospects for success, and then consider whether you want to handle any part of the appeal yourself or whether you prefer to hand it over to a professional.

TARGET APPEAL TIMETABLE FOR WRITTEN REPRESENTATIONS (ENGLAND/WALES)

Timing	Action
Start	• Full statement in support of appeal prepared. • Appeal forms completed and submitted.
3–4 days	• Appeal registered as valid. Start date given.
Week 2	• Council sends all details of its planning file to Planning Inspectorate.
Week 6	• Council submits its statement.
Week 9	• Final comments: both cases made by council and appellant.
Week 12	• Site visit made by appeal inspector.
Week 17	• Decision letter issued.

DIFFERENCES IN THE SYSTEMS IN SCOTLAND, WALES, NORTHERN IRELAND AND REPUBLIC OF IRELAND

The planning system is broadly similar across the UK and the Republic of Ireland. Local councils administer a development plan system that shapes policy, and a planning applications procedure. They all have full and outline planning permission. In the UK, once an application is decided, an appeal can be made against refusal. In the Republic of Ireland, a decision is notified to the applicant and to anyone who commented on the application. Any of those parties can appeal within a four-week period, which means an objector can also appeal against the granting of permission.

Permitted development rights

All householders have rights enabling extensions, garages and outbuildings to be built without the need for any planning application. Unfortunately these rights, known as 'permitted development rights', are subject to so many provisos, clauses, subclauses and qualifying criteria that it is often not at all obvious what you can and cannot do in any particular situation. General points include the following:

• Permitted development rights can be removed, partially or completely, by a condition attached to a planning permission (for example, councils attach such limiting conditions to planning permissions for rural conversions).
• Some permitted development rights are restricted if you are In a conservation area, AONB or National Park or if your house is listed or part of a terrace.
• Councils can remove permitted development rights from a whole area by means of what is known as an 'Article 4 Direction'. These directions are most commonly applied to conservation areas. Double-check with your council's Planning Department whether the rights are restricted in your particular location.

EXTENSIONS TO BUILDINGS

Rights to extend houses are based on the size of the 'original house', meaning the house as it was in 1948 or as it was built if constructed since 1948. Any later extensions, including conservatories, use up those rights (see box, opposite).

ALTERATIONS AND BUILDING IN YOUR GARDEN

There are numerous permitted development rights allowing home alterations and buildings in gardens (see box, opposite). Rights to put up garden buildings are surprisingly generous but the use must be associated

PERMITTED DEVELOPMENT RIGHTS FOR EXTENSIONS (ENGLAND/WALES)

Extensions can be built subject to these main limitations:

Aspect	Limitations
Size	• An extension can be up to 70 m^3 or 15 per cent of the volume of the original building whichever is greater – subject to a maximum of 115 m^3. • Terraced houses and houses in National Parks, AONB or conservation areas can be extended up to 50 m^3 or 10 per cent of the volume of the original building whichever is greater, subject to a maximum of 115 m^3. • There are no permitted development rights on listed buildings. • An extension cannot cover more than 50 per cent of the garden and grounds.
Location	• An extension cannot be built closer to a public road adjoining the property than the original house, or closer than 20 m to the road if the house is more than 20 m away from it.
Height	• An extension cannot be built higher than the highest part of the original house. If built within 2 m of a boundary, the height of the extension cannot exceed 4 m.

PERMITTED DEVELOPMENT RIGHTS FOR BUILDING IN YOUR GARDEN

Aspect	Limitations
Buildings, enclosures and swimming pools	• If larger than 10 m^3 must not be within 5 m of the house (or is treated as an extension). • Cannot cover more than 50 per cent of the garden and grounds. • Restricted to 10 m^3 in National Parks, AONB or conservation areas. • Cannot be built nearer a public road or footpath than the original house, or if house is further than 20 m away. • A ridged roof cannot be higher than 4 m. • A flat roof cannot be higher than 3 m.
Fences, walls and gates	• A fence or wall adjoining a public road used by vehicles cannot be higher than 1 m. In other cases it cannot be higher than 2 m. • A fence or wall cannot be erected around a listed building.
Hard surfacing	• No restrictions.
Access to an unclassified road	• No restrictions.

PERMITTED DEVELOPMENT RIGHTS FOR ALTERATIONS (ENGLAND/WALES)

Home improvements can be made subject to these main limitations:

Aspect	Limitations
Roof extensions	• Volume of a roof extension cannot exceed the extension limitations given on page 49. • On a terraced house the volume of a roof extension must not exceed 40 m³. • On all other houses the volume of a roof extension must not exceed 50 m³. • A roof extension cannot be higher than the highest part of an existing house. • No rights in National Parks, Norfolk and Suffolk Broads, AONB and conservation areas.
Dormer windows	• Not permitted in any roof slope facing a road.
Other windows	• No limitations.
Porches	• Cannot exceed 3 m² in size or 3 m in height, measured externally. • Cannot be situated closer than 2 m of a boundary with a public road.
Cladding	• Cladding with stone, artificial stone, timber, plastic or tiles not permitted in National Parks, Norfolk and Suffolk Broads, AONB and conservation areas.
Re-roofing	• No change permitted in the shape of the roof.
Painting	• No advertisements, directions or announcements can be included in the exterior decoration of a building.

Investigate any limitations on extending an existing building. If you carry out work that is not permitted you may be required to revert to the original at your own cost.

with the normal enjoyment of the house rather than, say, for business use or to provide additional self-contained accommodation. The rights only apply to what is known as the 'curtilage' of the house – normally its garden or grounds – so would not include an adjoining paddock. Permitted development rights give rise to many disputes. Consequently, be certain to check carefully whether you need to apply for permission for anything you plan to build.

NATIONAL PARKS, SITES OF SPECIAL SCIENTIFIC INTEREST AND AREAS OF OUTSTANDING NATURAL BEAUTY

Planning requirements in National Parks are restrictive in the interests of preserving the character of the area. Permitted development rights are restricted and policies impose strict constraints on both new buildings and extensions to existing ones. Sites of Special Scientific Interest (SSSI) contain features of geological importance and nature conservation. Additional restrictions are imposed by law and via local plan policies to ensure these features are protected.

Areas of Outstanding Natural Beauty (AONB), or National Scenic Areas in Scotland, are areas of countryside deemed particularly attractive and worthy of preservation. Local plan policies give them extra protection, the emphasis being on preserving the appearance of the countryside. In an AONB you need to take extra care with design and landscaping to ensure you meet this objective.

BUILDING IN A GREEN BELT

Green belts are areas of land around major cities that have been established to prevent the outward sprawl of buildings into the countryside. By no means all countryside is considered green belt and not all cities have a green belt. Policy aims to prevent most new building. Sometimes village centres are excluded from the green belt while, at other times, entire villages are 'washed over' by green belt policy.

If you are proposing to build in the green belt, check the local council's policies carefully to make sure your plans are realistic, because it is almost impossible to put up a new house in green belt countryside.

If you plan to develop a site in an AONB you should be aware that planning restrictions are likely to be more stringent than in other, undesignated, areas.

LISTED BUILDINGS

Listed buildings have architectural or historic interest and are given special protection. It is an offence to demolish or alter a listed building without Listed Building Consent. You can convert and extend a listed building, but high standards of design are essential. Note that, if you are building near a listed building, the possible affect of your new house on its setting will be taken into account.

BUILDING IN A CONSERVATION AREA

Conservation areas are specially designated parts of towns and villages where the historic character receives extra protection. Local plans contain policies on development within these areas. They require sensitive design for any new building, which has to preserve or enhance the character of the area. Conservation Area Consent is needed for demolition.

CATEGORIES OF LISTED BUILDINGS
England/Wales
Grade I The most exceptional buildings (2 per cent of all listed buildings)
Grade II* Particularly important buildings (4 per cent of all listed buildings)
Grade II Buildings of special interest (94 per cent of all listed buildings)

Scotland
Category A Nationally important buildings (7 per cent of all listed buildings)
Category B Locally important buildings (62 per cent of all listed buildings)
Category C Good buildings with elements of interest (31 per cent of all listed buildings)

Northern Ireland
Grade A
Grade B+
Grade B1
Grade B2
Grade C

Republic of Ireland
There is one grade of listed building: a structure placed on the Record of Protected Structures.

TREE PRESERVATION ORDERS

Tree preservation orders (TPOs) give a tree or group of trees protection from felling, damage, lopping or topping. Unauthorized work on a TPO tree is potentially a criminal offence, punishable by fines of up to £20,000. Always find out whether there are any protected trees on land you are planning to develop. You can do this by checking the council's records held by the Planning Department. Finding protected trees on your land does not automatically mean you would not be able to build, as planning permission can override a TPO. On the other hand, it does mean the council will look very carefully at whether building work will affect the trees and it could use the loss of trees or potential damage as reasons for refusal of planning permission.

HEDGEROW PROTECTION

Some hedgerows, excluding those around the curtilage of a house, are protected by the Hedgerow Regulations. If you are building in a former paddock or converting a rural building you might come across a protected hedge. The regulations are quite complex, particularly if you want to remove the hedge altogether. Most often, self-builders only want to create access through a hedge, which is allowed provided the new access is in substitution for an existing one, or where there is no other means of access, or where the only alternative would be prohibitively expensive.

LANDSCAPING ISSUES

Existing trees and hedges can provide a ready-made mature setting for a new house, potentially increasing its value. They can also provide vital screening between neighbouring properties, preserving privacy and preventing privacy from becoming an issue in a planning application. The screening effect of trees, hedges and bushes can enable greater flexibility in the size or design of a new house. New planting can also be designed to achieve or enhance all these benefits. Even if you are only making an outline planning application, it pays to show some landscaping on any illustrative layout plan you submit. Any layout plan accompanying a detailed application should always show landscaping and, where landscaping is an important part of the scheme, you should submit a full planting scheme with the application.

RURAL SITES

Some local plans have a specific policy allowing the infilling of small gaps in rows of houses in the countryside. Others specifically identify those parts of villages and hamlets where infilling is accepted. Some go so far as to describe the maximum size of gap that can be infilled. All such policies restrict infill to small gaps in otherwise built-up frontages or within groups of houses. Where there is no official policy, even genuine infills are likely to be refused by the council.

If you run, or are setting up, a full-time, profitable agricultural business (including horticulture), you might be able to build a house if it is absolutely necessary for the business to have someone living on-site full time. In the case of a new business, councils often grant temporary permission for a mobile home (three years), to ensure the business establishes and becomes viable. Only then would permission be granted for a permanent house. Size restrictions are imposed and over-elaborate or expensive designs discouraged.

Occasionally, you might come across land for sale with planning permission for a house subject to an agricultural occupancy condition or 'tie'. This means only those employed or formerly employed in agriculture, and their dependants, can live in the house when built. It is sometimes possible to get these ties lifted from existing houses, but you must prove there is no demand for agricultural dwellings in the area. This is usually done by marketing the property at a reduced price (which reflects the tie). Typically the property is marketed at a reduced price for 6 to 12 months. If no one comes forward to buy then there are reasonable prospects for getting the tie removed.

BROWNFIELD SITES AND URBAN PLANNING

'Brownfield' sites are currently favoured as locations for building new houses. These are usually urban areas of previously developed or used land. This means any land that has built structures on it (excluding agricultural buildings), but not open land in built-up areas, such as parks or playing fields. While acres of urban industrial dereliction clearly fall within the definition so, perhaps surprisingly, do everyday houses and their gardens. Building at higher densities is encouraged on such land so some houses with good-sized gardens can now get permission for a new house or two in that garden, even where such a scheme might have been rejected a few years ago.

Council policy now generally makes it easier to get planning permission on brownfield sites. Factories, warehouses, workshops and empty space over shops are all candidates for conversion and are supported by policies that advise flexibility in parking standards, areas of amenity space and on issues such as overlooking. Some buildings previously considered unsuitable for conversion to housing – perhaps because of a lack of available off-street parking – can now get permission. There is often scope for innovative design and the creation of new and interesting houses.

A common issue with the more stereotypical brownfield sites, such as old factories or petrol stations, is ground contamination. When you make a planning application for housing on potentially contaminated land the council usually asks for a site investigation to establish what is in the ground and, if it is contaminated, for details of measures to clear up the site. It is your responsibility to carry this out.

THE FUTURE

The planning system in the UK is subject to continuous review as politicians struggle to meet housing demands while trying to preserve the environment. Changes are underway and more are predicted. At the time of writing, the local plans system, for example, is being revised. Expect new jargon and some uncertainty when the new system takes over. Almost any aspect of the system could change, so always check your facts with a local-authority planning officer or an independent planning consultant.

How to purchase land

Buying a plot of land or a property to convert is usually no different from buying a house. What is important is that you get proper legal advice before committing yourself: a house that turns out not to suit your requirements can be sold on; a plot that cannot be built on for some reason could turn out to be worthless.

BUYING THROUGH PRIVATE TREATY

Most plots are sold by a process known as private treaty. This involves negotiations between vendors and purchasers prior to an offer being made and accepted. You usually make an offer through the vendor's estate agent. This can be done by phone, but it is sensible to follow up with confirmation by email or fax. The agent will always confirm your offer in writing. Offers are not legally binding until contracts are exchanged. Once your offer is accepted, your solicitor undertakes searches and enquiries before contract, to make sure there is good legal title to the property and that there are no threats to the value of your ownership. After exchange, the sale completes on a pre-determined date stated in the contract. The process usually takes from 8 to 12 weeks from acceptance of an offer through to completion, although both longer and shorter periods are common. If there are a number of offers for a property, estate agents sometimes hold an informal tender. This means they invite 'best and final' offers, to be submitted in writing by a given date. These offers are not legally binding until the exchange of contracts.

BUYING AT AUCTION

At an auction a binding contract is established when the hammer falls. You have to write out a cheque for the deposit there and then, and completion usually follows a month later. This means you and your solicitor have to undertake all your enquiries, site investigations and legal searches before the auction and without any certainty of successfully buying the property. Auctioneers sometimes assist by producing an information pack for buyers.

BUYING THROUGH FORMAL TENDER

Occasionally, plots are sold by formal tender – an offer that is binding once accepted by the vendor. Where a property is for sale by formal tender, the selling agent produces tender particulars and conditions of sale. These documents include the time and place for submitting your tender and should be passed to your solicitor immediately if you intend to participate. All your pre-purchase checks have to be carried out before you submit your tender. You will be informed whether you have been successful or not once all the tenders have been considered.

BUYING PROCEDURES

Private treaty	Auction	Tender
Investigations	Investigations	Investigations.
Negotiation	Offer/acceptance/ exchange of contracts	Submit tender
Offer	Completion	Acceptance/ exchange of contracts
Acceptance		Completion
Exchange of contracts		
Completion		

BUYING IN SCOTLAND

Scotland has a different legal system from the rest of Britain, so you must use a solicitor who is a member of the Scottish Law Society. When buying a plot, one of the first things you will notice is that, in most cases, offers are invited in excess of a given figure. ('Above' probably means a good 10 to 15 per cent above, although this varies with the state of the local market and the amount of interest in a particular plot.) When you are ready to make an offer, your solicitor puts it forward in writing. All being well, the vendor's solicitor then responds with a qualified acceptance. Your offer normally contains the price and entry date (which is when you want to take over the property) and can be made subject to a soil survey or receiving planning permission. A binding contract – the conclusion of missives – will not occur until the survey is complete or planning permission obtained. Once missives are concluded, your solicitor conducts the various searches to ensure there is good title to the land and your purchase moves forward to settlement (completion) on the entry date.

Buying property in Scotland is generally a quicker process than in the remainder of the UK. Make sure you carry out all your checks and homework on the plot before your offer goes in, as its acceptance is binding. That said, not all transactions are handled this way, and some proceed via offer and non-binding acceptance, followed later by exchange of contracts and completion.

BUYING IN THE REPUBLIC OF IRELAND

The Republic of Ireland has a similar system for buying property as the UK but Irish law is, again, different and you should consult an Irish solicitor on your purchase.

BUYING TACTICS

The most fundamental question when buying land or embarking on a conversion concerns planning permission.

If the property has full planning permission for exactly what you want to build, there is no problem. However, if the planning permission is either an outline or a detailed permission you want to change, you have to decide how much risk you are willing to take.

If planners, neighbours and councillors are happy with your scheme, you could afford to wait until completing your purchase to get planning permission. Such certainty is quite rare, however, and a better approach is to try to get your permission before the sale completes. You might achieve this by submitting your application as soon as your offer has been accepted. By the time you get to exchange of contracts, the application should be far enough advanced for you to see which way it is going. In a busy market, vendors will not want to delay exchange while your planning application goes through. On the other hand, if you do get permission before exchange, you are vulnerable to gazumping if your permission increases the value of the plot. One way out of this impasse is a conditional contract.

CONDITIONAL CONTRACTS

You might be unwilling to exchange contracts on a plot or building for conversion until you have got the planning permission you want, or until some problem, such as a restrictive covenant, right of access or drainage difficulty is resolved.

In these circumstances, the best way forward is to negotiate a conditional contract.

Put simply, you exchange contracts, with the contract specifying that you will not complete until and unless the condition has been satisfied. You might also want clauses inserted allowing for renegotiation or an appeal, if permission is refused.

WHAT ARE OPTIONS?

Options are used by developers to secure the right to buy land at a future date. A sum of money is paid to the landowner, who grants an option to the developer in exchange. The option lasts for an agreed period of years, during which the developer can buy if planning permission is granted. The developer is then responsible for promoting the site and obtaining planning permission. If successful, he buys the land at an agreed discount below its open-market value. The discount reflects the risk, the expense of getting permission and the up-front payment. Options are rarely bought on single plots but could be a suitable way forward for a landowner who has no interest in trying to get planning permission him- or herself.

HOW MUCH SHOULD I EXPECT TO PAY FOR LAND?

The price to pay for land is a compromise between what it is worth to you and what other people are prepared to pay for it. You can calculate a figure by establishing what the finished house would be worth and then subtracting all the costs involved in building the house. The amount left over is, theoretically, what you can afford to pay for the site, assuming you need the project to balance financially. This figure might be more or less than others would be prepared to pay, so you have to take into account the level of market prices for land in that area.

It used to be a general rule that one-third of the value of the finished house would be paid for the land. In recent years, in areas of short supply, 50 per cent of the anticipated value of the finished house is frequently paid. In less buoyant markets, the percentage might drop to 25 or even 20 per cent.

The first step in valuing a plot is to have an estate agent estimate the value of your house once complete. Get quotes from two or three agents. If local agents are selling a good volume of plots, ask their opinion on plot values and the percentage of finished value being paid. Fine-tune your figure to reflect any exceptional costs, such as special foundations or excavations to deal with a sloping site. You will have to pay some regard to the asking price and vendor's expectations but do not be tempted to overbid if those expectations seem unrealistically high.

VALUATION METHODS

- **Ask an estate agent** Agents who deal with plots regularly have a feel for the market and can give you helpful advice. House agents can also provide accurate estimates of what a house will be worth when built.
- **Comparable sales** The sale prices of comparable plots in the area will act as a good measure. Make sure they have planning permission for similar size/value house as yours.
- **Rule of thumb** Compare the sale price of a plot with the value of what can be built on it. For example, a plot that sells for £100,000 with potential for a £300,000 house shows a 33 per cent house-to-plot value. Compare several examples to arrive at an average percentage for your area. Use this as a rough guide to the value of any plot you find there.

FREQUENTLY ASKED QUESTIONS

Can I change planning permission on a site where it has already been granted?

Yes, you can change a permission, although an existing permission can set a benchmark for development of a site. Radical departure from that type of design can be difficult.

Do mobile homes need planning permission?

Yes. If they are to be used as a home, they are treated like any other house. They do not need permission if they are to be used as temporary, on-site accommodation during a self-build project. (One of the occupants must be fully engaged in the project either physically building or project managing.)

Are there special planning concessions for sunrooms and conservatories?

No, for planning purposes they are treated exactly the same as any other kind of extension.

Can I get round normal planning restrictions on building in the countryside by building underground?

Unlikely, although building underground removes most visual objections.

Can I use someone else's planning permission?

Yes. Planning permission attaches to the land and does not belong to the applicant. Note that copyright of the drawings will be owned by the building designer or, possibly, by the applicant, and you may need to negotiate in order to gain the copyright.

You see advertisements for plots at very low prices, usually where a whole field has been divided up. Are these plots a good buy for self-builders?

This is highly unlikely. These sorts of plot rarely have any foreseeable development potential at all – that is why they are so cheap.

Can any brownfield site be built on?

No, much depends on what and where it is. The emphasis on building on brownfield sites applies to urban areas.

If I keep a few animals and grow some vegetables, could I get permission for an agricultural dwelling?

Only genuine agricultural or horticultural businesses with a track record of profitable trading and a need for 24-hour supervision are likely to be granted permission.

Finding a plot checklist

FINDING LAND

- [ ] Register with estate agents
- [ ] Contact landowners such as councils and large companies direct
- [ ] Register with a land-finding agency
- [ ] Speak to property professionals
- [ ] Network with friends, family and work colleagues
- [ ] Speak to local builders and house developers
- [ ] Look up council planning records and local plans
- [ ] Study maps, search your area, identify and contact potential plot owners

ASSESSING POTENTIAL

- [ ] Check size of plot, boundaries and for obstacles
- [ ] Look for paths, pipes, cables and trees
- [ ] Ensure adequate access and services
- [ ] Get a soil survey and site survey and investigate flooding potential
- [ ] Check deeds for covenants, rights of way and other restrictions
- [ ] Study the planning permission, ensure it is valid
- [ ] Conversions – ascertain suitability of building and planning policies
- [ ] Renovation – find out whether planning permission is needed
- [ ] Demolition and re-build – make sure residential use is not abandoned and check local plan policy

PLANNING PERMISSION

- [ ] Always check time limit, conditions and plans
- [ ] Outline permission – find out what size/design will be approved
- [ ] Detailed permission – ascertain whether it can be changed
- [ ] When making an application, first check local plan policy and planning history
- [ ] Make your design consistent with the character and pattern of housing locally
- [ ] Avoid affecting neighbours' privacy
- [ ] Keep in touch with the planning officer
- [ ] Lobby neighbours and councillors
- [ ] Attend the planning committee
- [ ] If in doubt, take professional advice

BUYING A PLOT

- [ ] Never sign a contract until you know you can build what you want
- [ ] Consider a conditional contract if you need to get planning permission before buying
- [ ] Do not bid at auction or submit a tender unless you have completed all your checks
- [ ] To value land, talk to agents, look at comparables and calculate by rule of thumb

3 Design

It may come as a surprise to you that it can actually be quite difficult to build any house style that you like. One of the biggest factors determining what you can (and cannot) build is what your local authority will allow. While some are willing to let the modernist dream become a reality, others are quite conservative and will insist that house builders adhere to the local vernacular.

Finding an appropriate style

First and foremost, it is a good idea to get a clear idea of what is acceptable: you will find this information in the Planning Department's design guidelines for houses. These will tell you what style of house is likely to be accepted in your area. If you have additional ideas you can visit the Planning Department in person. Take along a map or plan showing where the site is, photographs of the site as it is and anything that shows the kind of development you are thinking of. You might even persuade the duty planning officer to make a site visit.

The prevailing style in the area will most likely be encouraged, and perhaps even enforced in sensitive locations. The more mixed the existing local styles, the more freedom you have and the greater the likelihood that something new to the area will be acceptable. Areas with a very homogeneous character will tend to need something that fits in, and only the most skilful, inspired designer will be able to produce something 'different' that works.

If you have some flexibility, consider that most house styles fall very broadly into two categories: the rational, symmetrical, formal 'classical' (such as the Georgian style); and the passionate, irrational, asymmetrical, decorative and informal 'romantic' (such as the gothic, medieval and much Victorian building).

If you are thinking of building in a historical style, find books that show plans of original examples to see how each era organized its homes. You can use these examples to influence your own plans.

THE SIZE AND SHAPE OF YOUR BUILD

Planners may be quite precise as to the size and shape of your new build and its position on the site.

One of the pleasures of self building lies in being able to mix different styles and materials. This detached house combines a timber-frame façade with stone walls.

A Georgian-style house will have clean, classical lines and multi-pane windows. Opt for authentic colour schemes inside for additional elegance.

This can also be dictated to some extent by its style, as specific styles suit certain plan forms and scales of building. The surrounding environment – the type of street and size of neighbouring houses – will also influence this. Bear in mind that size and complexity bring a higher cost.

ACCOMMODATION SCHEDULE

A key aspect of your house design will be determining how much living area you require and its position in the house. Start by drawing up an 'accommodation schedule': this is a list of the rooms you want and the size you want them to be. It can be developed to include other requirements such as favourite items of furniture and which rooms you want good views from.

MOOD BOARDS

To help you create a 'look' and 'feel' for each type of room or for the house as a whole it is a good idea to create a mood board. This is where you use a selection of cuttings from magazines, estate agents' brochures, photographs, off-cuts of fabric and so on, in order to build up a picture of how you want your home to be. For a couple or a family, this can be a useful way of communicating ideas to each other. You can create a mood board for each room of the house, and even the garden.

BUBBLE DIAGRAMS

A good starting point for working out the plan of a house is to draw up a 'bubble diagram', in which each room is represented by a circle. The sizes of the circles are relative to the sizes of the rooms. Those that need to be next to each other are connected by lines, and overlapping circles represent open-plan or multi-use areas. Outside influences, such as views, points of access and the passage of the sun over the site, should also feature. A bubble diagram can be cross-referenced to an accommodation schedule, which can, in turn, be cross-referenced to the mood boards.

Planning your rooms

Once you have an idea of the footprint, or floor area, of your house you can finalize the positioning of the rooms. This is something you might have been thinking about for some time, perhaps even

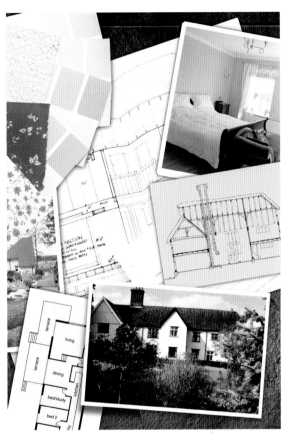

Mood boards are a great way to help you plan designs for your home. You can create a mood board for the whole house or each room of the house.

before buying the land. Now is the time to modify your plans to work with the shape and size of your proposed house. The important thing is to think long and hard about your requirements.

FUTURE PROOFING

You have a considerable amount of flexibility in determining the layout of the interior of your house, so try to design a house that will meet your needs for as long as you live there. For example, a young couple might want to plan for an extra bedroom or two if they intend to have children in the future; a couple approaching retirement might want to downsize eventually, planning for an occasional bedroom that can double up as a recreational room, or even for the removal of a partition wall at a future date. If you plan to sell the house on, make sure the layout is flexible enough for another buyer's needs.

QUESTIONS YOU SHOULD ASK YOURSELF

Use your current living arrangements to help you determine how much space you want in your new home and how you want various rooms to be arranged. Look at other people's houses to determine what you like or dislike about the internal layout of the rooms in different houses.

1 Is the current living space the ideal size, too big or too small? Work out the total area of your current home by measuring the length and width of all the rooms and adding them up. Make sure your new floor area is smaller or larger than this depending on your needs.

2 Does your current accommodation have rooms that are awkward in size or shape? Think about how you might change these rooms and incorporate the ideal into your new home.

3 Do you want more or fewer bedrooms than you currently have? Consider this not only in terms of your current family, but also in terms of future additions to the family and the likelihood of having guests.

4 Do you want more or fewer bathrooms than you currently have? Think about whether you want one big family bathroom, or all the bathrooms en suite.

5 Do you want to have an open-plan living space? For example, do you just want the kitchen and dining area to be open-plan, or would you prefer all the reception rooms to interlock?

6 Do you want to have an office? Think about where you might put it – on the ground floor, upstairs or above a garage, for example.

7 Do you want to have a room dedicated to a hobby? Consider whether this room needs to be in a quiet part of the house or near where all the action takes place.

8 Do you want a playroom for children that is separate from an adult reception room? Many self-builders have an adult reception room that they hardly ever use. You may decide this is a luxury that you can do without, to have more room elsewhere.

9 Do you want a conservatory or sunroom? Consider the financial benefits of including this in your original plans rather than adding one at a later date (see page 19).

10 Do you want the staircase and hall to be a feature or just functional? An imposing spiral staircase, for example, can be a very stunning feature of a house, but it may take up more room than a conventional staircase.

Now that you are building an entirely new home, you can afford to consider less conventional options. Perhaps you like the idea of an open-plan living space.

The fact that this kitchen has a sliding door means that it can be integral to the living space when preferred, but neatly tucked away at other times.

CALCULATING THE SIZE OF THE ROOMS IN YOUR HOUSE

Think carefully about each room in turn to make sure it is going to fit your requirements: it is possible to build rooms that are either too big or too small for your needs. Use your current living space to guide you: take note of areas you like and to work out their sizes and shapes by pacing them out. The ceiling height and size of windows can have a dramatic effect on the 'feel' of a room, and knowing that a door is usually 2 m high will help you to make a quick estimate. The key to self-building is to plan well in advance.

Use the chart below as a guide to the main rooms of a house, and the likely considerations you should make when thinking about size and shape.

REAL ROOM PLANNING

Once you have a rough idea of the size, basic components and layout of each room, sketch some initial designs for them. Measure the furniture to scale and draw them on graph paper. Cut out the various items and play around with these model replicas until you have a layout that works. Remember to allow space for doors opening and windows. Show your layout to your architect, house designer or builder to make sure it works and you haven't forgotten anything. If you want to know more about planning rooms, you will find a wealth of information in the many books published for architects and designers. You can contact the Royal Institute of British Architects' (RIBA) bookshop or the Building Centre for a list.

HOW MANY ROOMS DO YOU NEED?

By asking yourself the questions on the previous page you should be able to fill in this checklist.

Type of room	Number of rooms	Type of room	Number of rooms
Bedroom		Dining room	
Bathroom, main		Kitchen	
Bathroom, en-suite		Utility room/larder	
Shower room		Sports room	
Separate WC		Sauna/steamroom	
Living/play room		Other rooms	

Major constructions*	Number	Other constructions	Number
Garage		Shed	
Swimming pool		Folly	
Outbuildings			

Although optional, major constructions should be considered during the initial stages of designing a new house.

IN THE BEDROOM

- **Bed** If you have a specific bed in mind, measure it.
- **Hanging space** Do you want more or less hanging space than you currently have? Think about the convenience of a built-in wardrobe or separate dressing room.
- **Folded clothes and other items** Think about whether you want a chest of drawers, a tallboy or shelves/drawers as part of a built-in wardrobe/dressing room.
- **Bedside cupboard** Do you need one, or would a shelf near the bed do?
- **Full-length mirror** This may not be necessary for all the bedrooms, but is quite often forgotten. Remember to allow enough space for standing back to look check the clothes you are wearing.
- **Basin** A basin may be particularly useful in a guest room.
- **Chair/chaise longue** Think how you use your bedroom, and where you tend to leave your clothes at the end of the day. A chair could be useful, or a chaise longue, if you have the space.
- **Shelves or set of shelves** Consider whether you want to keep your books near you rather than in a separate room. They tend to get dusty in a bedroom so you may want to have a glass door or have just a small collection in the bedroom.
- **Lights** Think about where you want them, particularly if you want to have wall lights by the bed or a stand-up lamp that takes up floor space. Would you benefit from a separate daytime and night-time lighting system?

If you already have bedroom furniture that you know you want to keep, you can use this as a starting point for designing a bedroom in your new house.

Not only are mirrors useful for checking your
clothes, but they can also make a small bedroom
appear more spacious.

IN THE BATHROOM

- **Bath** If you want a bath think about the ideal size you want – for one person or two? Do you want it to have a shower attachment and screen/curtain? Do you want the taps at the end or in the centre of the bath?
- **Shower** Should this be integral to the bath or stand-alone? Think about the ideal size – it could double up as a mini steam room.
- **Basin** Is one enough? Should it be a small basin or a large one? Do you want 'his' and 'hers' basins?
- **Mirror** This typically goes above the basin.

- **Lights and electric razor points** Consider lights for both daytime and night-time activity.
- **Cupboard** Often overlooked, but usually essential. Planned well, a cupboard can be either very discreet or a feature of the room.
- **Heating** Consider the various heating options available – a heated towel rail or underfloor heating, for example.
- **Floor space** Think about the remaining area once everything is in place. Is there enough to space to move around in and to get dry in comfortably?

In many houses bathrooms are compromised, squeezed in where there is space, but with a new-build you can have the room where you want and at a size to suit you.

Consider the style of bathroom you want. If you intend
to have a roll-top bath and bold colour scheme, make
sure you design a space that can take it.

RECEPTION ROOMS: DINING ROOM, LIVING ROOM, PLAYROOM

- **Dining room table and chairs** Do you want a separate eating area or one that is part of the kitchen? How many people do you want to seat on a daily basis, and how would you cater for a big dinner party?
- **Sofa and chairs** How many people do you want to seat at any one time? How many seats should be facing the television? Consider the size of any sofas and armchairs carefully.
- **Occasional tables, coffee tables or cupboards** These provide useful surfaces for drinks and gadgets. Think about where you want to put them.
- **Shelves** You'll need these for books, magazines, DVDs, videos, CDs, cassettes, framed photos, ornaments and so on. Think about whether you want them in a cupboard or on show.
- **Television and cinema system** Where do you want this equipment? Consider having the screen on the wall.
- **Hi-fi system** Should the speakers be discreet or a big part of the impression of a room?
- **Children's toys** Do you want to allow for cupboard space here?

Think about how you can design a dining room to fit your lifestyle. The room above is well suited to both family breakfast time and evening meals with friends.

Consider the style of your house when planning your interior. A clean, contemporary living space will work well with many house designs.

KITCHEN, UTILITY ROOM AND PANTRY

- **Arrangement** Consider whether you want separate rooms, or one big open-plan space.
- **Dining table and chairs** Do you want to section off the eating area? How many people do you want to seat on a daily basis?
- **Cooker/oven** Consider how many you need – including grills and hobs – and think about what sizes they should be.
- **Sink** Should this be double or single? Think about how much draining space you will need and whether you want a built-in waste-disposal unit?
- **Cupboards** These can be freestanding units or built in. They can house integral dishwashers, washing machines and fridges. Consider carefully how many units you are going to need.
- **White goods** Think about a fridge and/or freezer, washing machine, dishwasher, tumble dryer. How big should they be, and where will they go?
- **Extras** You may need vegetable baskets, a waste bin, a towel rail, a wine rack and so on. Do you want them to be integral to units or visible to the eye?
- **Lighting and plug points** You'll need plug points and lights near the work surfaces. Lighting needs to be functional for food preparation, but consider adding mood lighting if you plan to eat in the same room.

Anyone with a passion for cooking and entertaining will know that a well-fitted kitchen is a must. Plenty of storage and the latest gadgets will be essential.

Even if you have a separate dining room, you may want to install a breakfast bar in your kitchen for informal eating or just as somewhere to sit and chat.

The oven, sink and fridge have been positioned to form a 'work triangle' in this kitchen. This means that all the cooking activities take place within one triangular area.

If you want a sleek, minimalist kitchen, you need to consider your storage options carefully. A pantry is a great idea for tidying away both food and appliances.

Using professionals

In theory you can design your house but there are risks. Development is highly regulated and bureaucratic, and to get through it without professional guidance would require a great deal of time. There are many issues to consider, many of which are not at all obvious to the lay person: land law, party-wall legislation, rights and easements, the legalities of highways, sewers and local enactments, getting services connected and more. Furthermore, your design has to comply with building regulation requirements, for example, for light, ventilation, disabled access and means of fire escape. Without really knowing the full legal situation you might design a house that cannot be built.

By hiring a professional, not only will you save time in establishing a comprehensive, working design for your build, but you will also avoid the likely pitfalls that many self-builders encounter. It is also worth noting that town planners will often reject an amateur submission if it is incomplete.

You have the choice of using three different kinds of professionals:
• an architect
• a technologist
• an architectural designer.

ARCHITECTS

The title 'architect' is reserved, by law, for those registered with the Architects Registration Board (ARB). Members will have completed a long training, usually involving five years of study (predominantly in design), two of apprenticeship, and examinations in law and contracts. The board regulates and disciplines members of the profession for the Government. An architect may or may not be a member of the Royal Institute of British Architects (RIBA), the Royal Incorporation of Architects in Scotland (RIAS), the Royal Society of Ulster Architects (RSUA) or the Royal Society of Architects in Wales (RSAW). These organizations promote the profession and support their members, as well as regulating them. An architect is trained primarily in design, but also in technical, legal and contractual issues, and is presented as being able to 'manage' – or, more precisely, and to avoid confusion with project management, 'administer' – a project from start to finish.

Architects' plans can be unfamiliar to a novice and it pays to go through them in detail before committing to a design. This will avoid misunderstandings later on.

TECHNOLOGISTS

A 'technologist' is a member of the British Institute of Architectural Technologists (BIAT). Design training is limited compared to that of an architect, but technical and managerial training is more thorough. Both architects and technologists are reliable in a technical, legal and contractual sense. Technologists are often presented as having a complementary role to an architect, and the key point is the difference in emphasis in their training. A substantial part of the five-year, college-based education of an architect involves the study of design, while it forms just one or two modules of a technologist's entire training.

Anyone with a professional qualification, or who presents themselves as an expert, is assumed by law to take on a 'duty of care' to you in respect of the skills they offer. Be aware of the limitations of the skills of each type of designer, and do not assume that he or she will take responsibility for any matters in which he/she is not fully qualified.

CONTRACTS

People are not always very clear about what they expect from a professional. Some self-builders want help 'managing' their build, and professional services are normally geared to all the building work being done by one, competent, builder. If you intend to employ subcontractors, the role of the professional and the paperwork involved are quite different. If you intend to be your own builder and lean on professionals to help, you may increase their workload considerably. This needs to be made clear right from the start.

Architects tend to use standard forms of contract with their clients. For large projects, they use a Standard Form of Agreement for the Appointment of an Architect. For smaller projects, a Small Works contract can be used in conjunction with a Joint Contracts Tribunal (JCT) contract, which is used for agreeing work between the builder and self-builder. The architect's contracts will be provided by the RIBA member or can be sought directly from RIBA. A JCT contract can be bought from high-street newsagents or from the RIBA bookshop. Technologists and technicians use the similar 'Confirmation of Instructions' and 'Conditions of Engagement'. These contracts will be provided by a BIAT member or you can get them from BIAT.

All these documents set out precisely what the professional's, and your, responsibilities are, and there are 'options' to adapt the service to your needs. Although these standard forms appear complex, your professional should go through them with you and explain everything. They need some adaptation to suit the self-builder, which forces you to decide and agree on the responsibilities of each individual, and the small print outlines ways of resolving any disputes that might arise.

HOW MUCH WILL A PROFESSIONAL COST?

Fees vary widely according to location in the country and between urban and rural practices. The only way to gauge the local market is to obtain estimates from several designers.

PROFESSIONAL TRAINING

Designer	Design training	Technical training	Personal Indemnity Insurance (PII)	Professional body
Architect	Extensive – explores historical, psychological and cultural issues	Limited – generally assumed to be developed through experience	Mandatory	Royal Institute of British Architects (RIBA); Royal Incorporation of Architects in Scotland (RIAS); Royal Society of Ulster Architects (RSUA); Royal Society of Architects in Wales (RSAW)
Technologist	Limited – required only to demonstrate competance	Extensive – Thorough knowledge of principles and technologies	Mandatory	British Institute of Architectural Technologists (BIAT)

NB This table refers to training, and is not intended as a judgement on design ability.

Technologists used to be cheaper than architects but you may find that the gap has closed. You may also find a local technician or designer to do drawings and submissions for a few hundred pounds, but make sure you know from the outset what you are, and are not, getting for your money.

HOW SHOULD I CHOOSE A PROFESSIONAL?

There are a number of factors to consider:

- **Skills** If you have technical, design or legal experience, you can choose a professional whose skills 'dovetail' with your own. Generally, the bulk of design work is for commercial clients who have little interest in materials or fine detail. As a private client you are at the most demanding end of the scale. Consequently you need a designer who is aware of this and the amount of work he/she will need to do to satisfy you.
- **Location** Getting a professional to visit a remote site can be expensive, so it makes sense for you to find a local designer, who may also have a rapport with local planners and an understanding of what will get approved. On the other hand, if you are particularly keen on a designer's work, he or she can design a house for you with limited site visits. When it comes to the building stage of the project, there are certain things a designer can do remotely (such as resolving problems by use of digital camera, email and fax) and cannot do (such as quality control or recommending builders).

- **Tastes** Designers have their own preferences of style, and this should be clear from examples of their work. Some may cover a range of styles, showing that they have a more flexible approach. The main point is to find common ground. Look at previous work. If you like what you see, it is more likely that the designer will feel comfortable designing something similar for you.
- **Personalities** The more 'personalized' your design, the more your designer will have to know you and the more you will get to know him or her. The building industry, with all its stresses, satisfactions and disappointments, brings out the best and the worst in people. The more sympathetic and supportive the relationships are, the more efficient the whole experience will be. The best results come from projects where people work hard to reach what they see as a worthwhile goal: if people fall out, this becomes quite impossible.
- **Self-build experience** Self-building can place additional demands on professionals, and this is particularly the case with the more 'hands-on' self-builder, as the professional tends to fill the gaps left by the competent builder. A designer with self-build experience will not only be able to judge how much extra help you might need, but also how and when to offer it.
- **Availability** At busy times, you can expect a good designer to have a full workload. Do not be tempted to ask someone to 'squeeze you in'. It is difficult for an enthusiastic designer to turn down an interesting commission, but he or she may find it impossible to give your project the attention it deserves. This will result in unresolved design issues and/or delays in the running of the project.
- **References** Satisfied past clients will be happy to discuss their experiences.

WHERE CAN I FIND A DESIGNER?

Architects can be found through the professional institutes listed on page 71. Each has an advice line for clients. RIBA has a searchable database on its website; the RIAS is developing one; the RSUA and the RSAW sites link to the RIBA database, and the

---

AMBIGUITY OF THE FREE MARKET

In a free market, professional bodies will not be drawn on what their members should charge. RIBA used to issue guidance on fees, expressed as a percentage of the value of the building work, but they have been obliged to stop doing this. Instead they will be publishing guidance on how fees can be worked out. The suggested methods follow, and there is no reason why these could not apply to all types of designer.

- As a lump sum.
- As a rate for time spent.
- As a percentage of the cost of the building work.
- As a percentage of the value added to the property.

NB *The last method is a fairly new concept, and arose from architects wanting to show that good design leads to enhanced value.*

RSUA has a shortlist of practices not on the RIBA database. The RIBA database can be searched by location, build budget and particular interests such as low-energy. Architects can post pictures and descriptions of their work on this site, or have links to their own websites.

Technologists can be found through BIAT. Their website also has a database. The Association of Self-build Architects (ASBA) has a list of architects experienced in self-build, although the list is not comprehensive. Other leads are personal recommendations, searching the Internet (there are many architects with good websites) and telephone directories.

HOW SHOULD I INTERVIEW A PROFESSIONAL?

There are two possibilities, each with advantages and disadvantages. A meeting at the proposed site could lead to some early inspiration and a quick indication as to whether the designer's ideas are in tune with yours. However, not all designers come up with a concept so quickly, and this can lead to embarrassment and a sense of disappointment. A conscientious designer is more likely to spend the time looking for prosaic things like drains, as he/she is obliged to see if the site is feasible before starting on a design.

Meeting at the designer's office will allow him or her to show you any previous work, and perhaps sketch out some ideas, in comfort. The office will also tell you more about the designer: if you feel comfortable in his or her working environment, this is a good indication that you will get along. However, you may want to test his or her response to your site, and many designers will be reluctant to do so without visiting it.

Wherever you meet, it is courteous to offer to pay for a professional's time. The designer will be anxious to impress you, and is likely to offer reasonable advice at this early stage. If paid, he or she is likely to be more generous with his or her time. Calculated in the context of the sums you will be parting with over the course of the project, a few hundred pounds spent on testing and forging the relationships on which the success and enjoyment of the project rest, will be money well spent.

At this stage it is worth establishing who will actually be doing the work. In larger practices work tends to get delegated and you may find you interviewed a director but get your house designed by a colleague or junior. Although there is nothing wrong with this in principle, and directors are obliged to oversee their employees' work, it may not feel like the personal service you were expecting.

HOW SHOULD I WORK WITH A PROFESSIONAL?

Employing a professional doesn't mean giving up control. Many clients bring ideas of their own expressed in all manner of ways from sketched outlines to computer-designed models. A good designer will know and respect the problems of presenting ideas visually and will encourage other means, such as mood boards and descriptions, to tease out what is of fundamental importance to you. Don't be embarrassed to present your own drawings – even if not accomplished, they may carry a good deal of useful information.

It is essential to brief your designer thoroughly, even if some of your ideas are still quite vague or intuitive. Basic issues of budget and the accommodation schedule should be set out quite clearly from the outset. Although they may be adjusted later on, you should be aware that making radical changes that require a complete redesign would be costly. You also need to be very clear about the level of quality you want, as this affects the size of house you can achieve within your budget. Complex information-technology systems or top-end kitchens are examples of items that will significantly affect the budget.

DEVELOPING THE DESIGN

Do not be dismayed if your designer wants to move in a different direction: a good designer should both welcome and challenge what is proposed. You might think you know what you want, but your designer will see things from many different angles. He or she may challenge your assumptions, and will have the technical skill to be realistic about what can be achieved. He or she will also spot idiosyncrasies, inefficiencies and unintended consequences of your ideas.

Frequently, people employ a designer and end up with a house that is quite unlike their preconceptions, and yet they are happier with it than with their original idea.

This is where you discover how important it is for you and your designer to be sympathetic in character. Some of the best designs result from a long process of putting together and taking apart, weighing-up and compromise. Ideas are developed and resolved through dialogue and the open and frank exchange of ideas and opinions. Some of the best designs come out of a robust exchange of ideas at the end of which neither side takes sole ownership, but both feel satisfied with the result.

Good designers seek satisfaction in what they do, and need to love the end product to make their efforts worthwhile. They may resist having to design outside their own tastes and could put up a surprising fight against a request to alter a window here or change the use of a room there, primarily because it has a knock-on effect on the whole design. Often they are right, and there are many stories of people being glad, on completion of the build, that they listened to their designers.

FINISHING THE DESIGN

If you and your designer are agreed on a strong concept your design can fall into place very rapidly. Progress will be slower if you do not respond significantly, either positively or negatively, to a designer's proposals, because he or she will have no idea where to go next. Progress can also be held up by the infinite number of design solutions there are and by the fact that your own requirements develop and change. For these reasons, it is important to have some real deadlines. Starting the building work on a particular date can be the best control: with your designer, develop a programme that covers the whole project. You may be surprised to learn how much time should be allowed, not only for design, but for bureaucracy, preparation of information, tendering and for builders' already busy schedules.

CAN I USE A 'PACKAGE' COMPANY?

There are many design and package companies tailoring products to the self-builder. They often have a team of designers – architects, technologists or technicians – who can either modify a standard design from one of their brochures or help you design a house from scratch. The level of service depends on the company concerned and, of course, the price you are willing to pay. This one-stop-shop service will suit those who would rather be treated as a 'customer', with a single supplier taking full responsibility. Often these companies can help with potentially stressful or contentious issues, such as sorting out planning permission or providing a list of potential builders and other subcontractors.

Architects' plans

Once a design has been agreed upon between client and professional, the designer will draw out the final concept in a plan. The first draft is extremely unlikely to be the final one: designers develop their drawings throughout a project, adding to and refining the information on them, and generally making them appropriate for their purpose at each stage.

Architect's plans usually go through the following four stages:
- **Sketch layouts** The initial design stage.
- **Planning** Showing the information necessary for planning applications.
- **Building regulations** Showing more technical detail that demonstrates compliance with the regulations.
- **Construction** Often the same drawings are used as for Building Regulations, but they are now cross-referenced to finer details and specifications to show the builder exactly what is required.

Some designers write detailed annotation on the drawings, while others prefer to keep them relatively clear. The latter use short descriptions that are cross-referenced to larger-scale drawings and written material on separate documents.

SITE PLAN

This plan shows the whole of the site and as much of the neighbouring sites, houses, roads and other features as are necessary to show how they all relate together. Generally, underground drainage and access arrangements are shown because these are outside the building.

1 North point: From this you can see which rooms, or parts of the garden, are in sun at various times and which are overshadowed by the house, trees and adjacent houses.

2 Designer's name.

3 Client's name.

4 Project title.

5 Drawing title.

6 Reference number: Usually set out as project number/drawing number/revision letter. Many drawings may be produced as a project progresses. It is vitally important that each has a unique number so that it can be referred to easily and unambiguously. Whenever a drawing is altered and re-issued the 'revision letter' – starting at 'A' – is added. This drawing is now referred to as '088/09 revision A' or simply '088/09A'. The previous version (088/9) will be withdrawn and filed away so that it is not used in error.

7 Scale: This indicates the size of objects in the drawing relative to the real world. An architect's scale rule has all the commonly used scales. (Note that when drawings are printed they can change size a little, so dimensions measured off the plan are not accurate.)

8 Description of revision.

9 Site boundary: Usually shown as a dot–dash line.

10 Drainage: Lines with arrows show the direction of flow and circles denote inspection chambers. Manholes are usually indicated by rectangles, although some designers use dotted lines.

11 Adjacent road and point of access to the site.

12 Trees: These are shown in various ways, but the size should be about the same as the actual spread of the branches. Sometimes the height of existing trees is given.

13 Existing adjacent buildings.

14 The proposed house: Shown as an aerial view of the roof, with the footprint (the outline of the walls) in a dotted line.

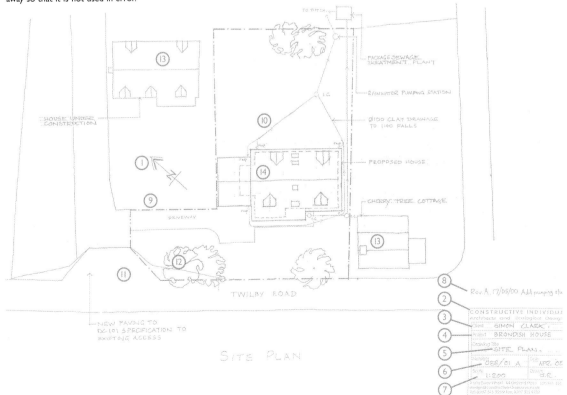

GROUND-FLOOR PLAN

This plan shows the layout of the ground floor. The drawing may show adjacent areas of landscaping such as paths and walls, but its primary purpose is to show the layout of the house.

I Title of each room.

2 Staircase: Usually an arrow shows the 'up' direction, although this is not universal. The stair is conventionally shown 'cut off' as it rises towards the upper floor.

3 Walls: The way these are drawn indicates how they are built. In this case, parallel lines show a timber-frame rainscreen wall, and diagonal strokes within a thin border indicate rendered blockwork. Another common convention is that paired diagonal strokes indicate brickwork. (The annotations refer to details on a separate drawing.)

4 Windows: These are often numbered with cross-references to a separate window schedule, which lists the windows and describes them in detail, usually with a drawing of each one. The manner of opening is shown on the elevation drawing (see page 78).

5 Doors: Numbered with cross-references to a separate schedule. They are drawn in the 'open' position, with the sweep of the door shown by either an arc or a diagonal line.

6 Kitchen layout: The basic layout is shown in the ground-floor plan – enough to locate the main appliances.

7 Rainwater pipe (rwp).

8 Soil vent pipe (svp): This is the main vertical drainage pipe, which also ventilates the drainage system.

9 Furniture: Designers may indicate furniture, to show how the room might be used. Mark in your own furniture to check that it fits and that you can use the rooms as intended.

10 Appliances: Designers use simplified outlines of appliances and lines with arrows to show drainage. These are indicative only – site tradesmen may work out the best layout for the drainage.

II Dotted lines are commonly used for showing things above or hidden below – in this case, the line of the roof.

12 Cross-reference to structural engineer's detail: Where necessary, a structural engineer will provide details and calculations to show that the building is structurally sound.

13 Access: The annotations and dimensions here show compliance with regulations for access and sanitary provision for the disabled.

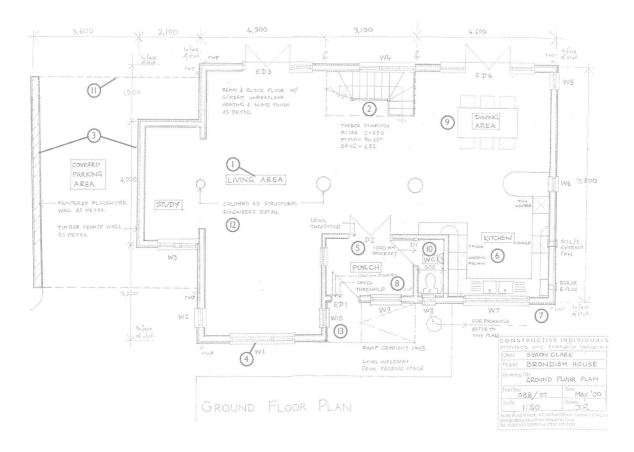

GROUND FLOOR PLAN

FIRST-FLOOR PLAN

This shows the layout of the first-floor rooms. The example also has some structural information.

1 Joist direction: This symbol indicates the direction in which the timber floor joists run. (This information may be shown on structural engineer's drawings instead.)

2 Beams: The dot-dash line shows where a structural beam runs in the floor.

3 Bathroom appliances: These are conventional symbols for a shower, lavatory pan (wc), wash basin (wb) and bath.

4 Roof ridge: The ridge is usually shown by a triple line.

5 Gutters: The annotations show that the gutter and rainwater pipe are adequate for the most severe storm likely to occur, as required by Building Regulations.

6 Roof fall: These broad arrows are often used to show the down slope of the roof.

7 Dimensions: These are often added for the builder's benefit. They do not always indicate the room sizes. Sometimes they are taken to the structure, ignoring non-structural and finishing layers for example. This makes the structure easier to set out for the builder, but they can be misleading as regards room sizes. They are almost always given in millimetres or metres.

8 'Equal' dimensions: Where dimensions are equal, rather than writing the figures out again, the designer will draw a stroke across each identical dimension line. If there is more than one set of equal dimensions, the first set will have one stroke on each dimension line, the next will have two, and so on.

9 Staircase.

10 Room titles: Bedrooms are numbered, which is less ambiguous than labelling them 'front bedroom', 'children's bedroom' and so on.

11 Void: Crossed lines are used here to indicate a 'void' or opening through to the floor below. They are also used to indicate gaps that will be filled later on (by kitchen appliances, for example).

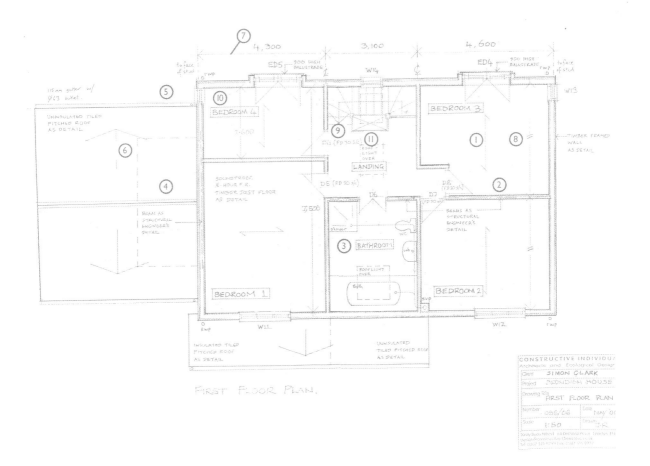

FRONT ELEVATION

The 'elevations' are views of the house from outside. They show finishing materials and vertical dimensions such as window heights.

1 Finishes: The annotation refers to more detailed information elsewhere.

2 Windows: The dot–dash lines show the manner of opening, with the lines meeting at the edge where the hinges are. This window has two side-hung casements and a central fixed pane. W8 is top-hung. In rare instances this convention is reversed, so it's best to double-check.

3 Soil vent pipe (svp): Seen here emerging through the roof. The dimension is critical in order to comply with regulations covering the proximity of soil vent pipes to openings into the building.

4 Rainwater pipe (rwp).

5 First-floor level (FFL): Marked on the elevation as a dot–dash line. This allows you to calculate the height of the windows from the floor.

6 Ground-floor level (GFL).

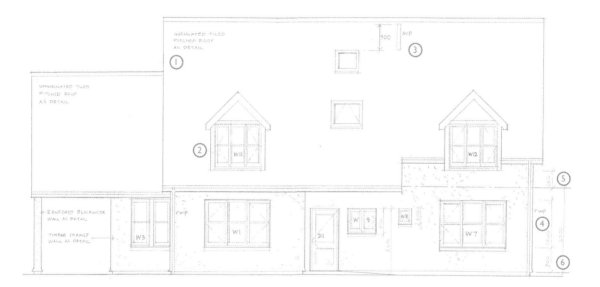

SOUTH WEST ELEVATION

Non-structural fittings and details

These fittings and details on the inside of a house include internal and external doors, door furniture (latches, bolts, handles, hinges), windows, skirting boards, cornices, architraves and stairs.

GENERIC STYLES

Although fixtures and fittings are functional, they must also be aesthetically pleasing and in keeping with the chosen style of your house. They are an extension of the house exterior and provide a good starting point for the interior design of your home. For example, if you are renovating a cottage, pieces bought from a reclamation yard will restore the building's original character.

Door style

If you are renovating a house, try to match any new doors you need with existing ones by visiting a reclamation yard or using a local joiner. If you are converting a stable block, you could install an old stable door, while in a cottage, ledge-and-brace doors are true to style.

Reclaimed doors suit cottage-style new builds as well, or you can buy ones made from distressed or reclaimed wood.

For an authentic-looking Georgian front door, paint it black, dark green or blue and fit a fanlight above it. Inside, six-panel timber doors can be fitted.

For an Edwardian house consider an external door with panes of coloured glass and use four-panel doors inside, perhaps with fingerplates made from porcelain or brass.

If you are building a Scandinavian home, look for a more contemporary feel. This style focuses on natural materials and clean finishes, so choose stripped wooden doors in light colours. Sliding patio doors, which can be pulled back to create a smooth transition between indoor and outdoor living, are also suited to this style of house.

This modern, coastal development maximizes on its sea views and is almost fully glazed to the rear, with large double doors out from the living room into the garden.

Window style

Traditional cottages tended to have few windows, some of which did not open. If you are renovating a cottage, you might want to consider dropping the window sills to make the openings bigger. If you want to install more windows in a cottage, particularly dormer windows, you must check with your local authority's Planning Department first, as there may be restrictions.

Very early cottage windows simply held a lattice of twigs for keeping out the rain. Tudor windows were also latticed – made up of small panes of glass held together by lead strips – and encased in a wooden frame that opened outwards. If this is the look you want, you can buy replica latticed casement windows with timber or metal frames, Traditionally, windows were set flush against the external wall so that the wall's thickness created a natural recess and an ideal place to sit. You may want to consider integrating a window seat into your design.

Sash windows, which slide up and down, are a feature of Georgian homes. Originally, internal shutters folded across the window at night to keep the house warm and dark. Bay or bow windows in the front rooms are typical of an Edwardian semi-detached house. In a contemporary Scandinavian home, windows tend to be large with big panes, and in the main living area they often reach from the floor to the ceiling.

A large Edwardian-style bay window will bring more light into a room and can be an attractive feature of the external appearance of the house.

Staircases

For a simple cottage, you should look for a timber staircase with a plain newel post and a simple balustrade. Farmhouse cottages often had a ladder staircase with slatted steps leading from the living space to the sleeping area above. Larger Georgian homes tended to be very grand and this is often reflected in Georgian new builds, where the staircase leads up to a galleried landing. If you have the space, you might want to add a gentle curve to the flight, and a wrought iron balustrade.

During the Arts and Crafts period, the emphasis was on hand-made, long-lasting goods. Staircases tended to have plain balusters fitted closely together and the treads were plain timber or painted white with the steps left bare. Heart shapes crafted into wood are very symbolic of this period and this was often the most extravagant touch to a plain set of stairs. Straight-flight stairs, leading up from a narrow entrance hallway, were common in Edwardian houses, typically with heavy, dark wood and square newel posts, capped with a decorative ball or something similar.

Architraves and cornicing

An architrave is the moulding around a door or window. In cottages and Tudor houses, timber lintels were built into the fabric of the house (above the window- and doorframes) to support the roof rafters.

Today, exposed lintels remain a common feature in new builds where they have the same function, and can be made using reclaimed wood instead of new timber. Ornate cornicing, between the top of the wall and the ceiling, was a feature of Georgian and Edwardian houses and it was often used to make an entrance look grander. Originally, cornicing was made of thick plasterwork or carved timber and was sometimes used as a shelf for displaying fancy crockery.

A spiral staircase will make a dramatic statement if it is positioned well. You will find that there are a variety of options available.

Simple timber staircases suit many house styles – from timber-frame to Edwardian cottage. The balustrade can be left natural or painted to suit the look you want.

PERIOD STYLES AND THEIR CHARACTERISTICS

Period style	Dates	Characteristics
Tudor house	15th –16th century	• Simplicity: wood panelling or bare stone walls with tapestries. Exposed timber beams. Oak furniture and fittings.
Georgian house	18th century	• Elegance and sometimes opulence. • Classical style: influenced by art and architecture in Europe.
Arts and Crafts house	Late 19th century	• Long-lasting, hand-made goods. • Oak furniture, tapestries, embroidery. • William Morris textiles.
Edwardian house	Early 20th century	• Less clutter and an emphasis on light. • A period of revival; a mixed style of medieval, Georgian and Arts and Crafts.
Scandinavian house	Present day	• Spacious and open plan. • Clean-cut, unfussy lines. • Emphasis on wood and neutral colours.

PLANNING AND LISTED BUILDING CONSENTS

Strict rules apply to how you can change the appearance of a listed building. If you need to replace some or all of the windows and doors in a listed building, you must seek consent before you begin the work. Rules apply to the inside of a building too: for example, you may not be able to remove an original staircase or alter the cornicing to carry out electrical work. Restrictions can also apply to new builds and buildings that are not listed. It all depends where you are building and the style of house. Planning departments tend to favour houses that stay in keeping with the buildings in the local area and unusual fixtures and fittings may be rejected.

Hard landscaping

The spaces outside your house are no less important than those inside. Done well, they can work as an extension of the space inside. Clever designers make the most of this by choosing external finishes with as much care as internal finishes, blurring the distinction between inside and outside even further.

Quite often, the design of a house will take up a vast majority of the budget, leaving little for developing of the rest of the site, yet poor hard landscaping can spoil the effect of a good-quality house. When it matches the house in terms of scale, style and quality, however, the total effect becomes more than the sum of its parts. Not only that, but if you carry out the hard landscaping at the same time

---

DOS AND DON'TS

Do stick to the same theme throughout the house.

Do use reclaimed materials where you can in a renovation or conversion project.

Do fit period copies in a new build but don't buy cheap imitation pieces.

Don't be heavy-handed: You want your interiors to be a gentle reflection of the exterior, not a tacky copy of the original.

Do have a look at original period houses to get ideas for your new build.

as you build the main house you can claim back the VAT (see pages 16–19).

PARTY-FENCE WALLS

When building walls on your site boundaries with other private owners, you have two options: you can build the wall on your side of the boundary and retain ownership of it; or you can agree with the neighbour to build the wall straddling the boundary, share the cost, and create a 'party-fence wall'. With the second option you both have rights to attach things to the party-fence wall, such as a lean-to building, and you both have rights to enter each other's property to maintain the wall. Consult a party-wall surveyor for advice and to formalize the paperwork.

ERECTING WALLS AND FENCES

Walls and fences can define spaces, form edges to changes in level, provide security and privacy, and create shelter. Walls create boundaries and spaces with a sense of permanence. They can soak up heat,

creating a warm 'microclimate' either for your enjoyment or to suit certain types of plant. Most types of brick can be used, although frost-resistant bricks should form at least the capping (top) course and the lowest courses, because these tend to stay damp and can easily be damaged by frost. Alternatively a tile, stone or concrete capping can be used. A damp-proof course will help to control rising damp and avoid green growth.

The depth of the foundation is dependent on the soil. Ideally it should be deep enough to avoid movement resulting from soil heave or freezing (450 mm in non-cohesive soils and 900 mm in cohesive soils, see page 108). In reality, shallower foundations are used because a little movement in a garden wall is not a big problem. Long garden walls tend to have movement joints so that whole sections move independently rather than cracking. A better approach might be to lay the bricks in pure lime mortar, which will allow a bit of movement and easy maintenance. This might also be a good approach for walls that tend to get damaged, such as

Successful hard landscaping lies in choosing materials and textures that work well with the fabric of the house you have built.

those at vehicle entrances. A retaining wall needs to be built properly if it is not to tip over, crack or suffer from dampness. This is achieved by providing drainage behind it, an adequate foundation and using the correct materials.

When it comes to erecting a timber fence, there is a tremendous variety of styles to choose from. Here are the most popular:

- **Post and rail** This can have a very 'rural' appearance and is commonly used with agricultural five-bar gates.
- **Palisade** Normally 900 mm or 1200 mm high, this consists of boards, shaped at the top, and with gaps between.
- **Boarded** This kind includes the 'horizontal lap' (horizontal overlapping boards), the 'waney-edge' (horizontal overlapping boards that have one wavy edge following the shape of the tree from which they were cut), and 'close boarded' (vertical overlapping boards). They are supplied in standard panels 900 mm, 1200 mm, 1500 mm or 1800 mm high, or can be fixed to posts and rails to give a continuous appearance.

Think about creating decks for eating, patios for sun-lounging and play areas for children – all completely separate from the rest of the garden.

The careful use of trellises and pergolas can divide a large external space into a series of separate outdoor 'rooms'.

• **Trellis** This is made from strips of wood attached to a frame so that they form an open mesh.

Fences incorporating willow, bamboo and heather used to be rural craft products but have more recently become easily available in panels from mainstream merchants. Willow fence panels are sometimes referred to as hurdles.

There are many metal fencing systems that look more appropriate in front of public buildings or around school grounds. More domestic are the wrought iron palings or decorative panels that come in standard-size panels or bespoke (made to measure) at a higher price. Mild steel fencing can be made to your own design and need not be very expensive.

A fence or wall erected to keep intruders out should be strong and difficult to climb. Ways of enhancing security are to avoid horizontal rails that might provide a foothold, to add a weak trellis to the top (this breaks noisily) and to encourage spiky plant growth. Most potential intruders will be put off by the risk of injury. Make sure that any sharp

If you are planning on building a driveway, make sure you specify materials that are robust enough to take the wear and that it is built to fall for drainage.

deterrents are visible – you have a duty of care to avoid injury to intruders. Also beware of creating dark areas where potential intruders can hide.

If graffiti is a problem in your area, consider open-boarded fencing, mesh or railings, perhaps shrouded in a hedge for more privacy. A close-boarded fence can at least be repainted, unlike brickwork, which is extremely difficult to clean. Graffiti-resistant paints are available.

LAYING PAVING
Any hardwearing, strong, moisture- and frost-resistant material can be used for paving, as long as it does not get slippery when wet. Even quite weak or thin materials can be fixed down to a concrete slab, or you can make a mosaic.

The more attractive paving finishes are generally more expensive, while the more affordable new paving, such as tarmac, interlocking blocks or concrete slabs, look uncompromisingly modern. It is therefore difficult to pave in character with an old-style building and on a limited budget. The cheaper artificial or stone-effect paving is seldom convincing owing to the regular size of the slabs. Real stone flags are, of course, expensive. In between is a range of clay and coloured concrete paving which, if selected carefully, can be very attractive, and will mellow with time. For an interesting 'green' option, look at recycled rubber paving. Gravel is good for security (you can hear people coming) but can get messy if not very well laid and maintained.

Subbases
All paving needs a stable base. This is done by removing topsoil and, if necessary, any weak or unstable soil beneath, and the level brought up again with well-compacted hardcore, referred to as the subbase, usually at least 150 mm thick. Driveways need to take considerable loads and require thicker layers of support under the paving if they are not to sink under the weight. Specify the subbase to take the heaviest vehicle you expect.

Most paving is laid on 50 mm of sharp sand on top of the hardcore. This sand has to be quite flat and laid to the right falls (see page 86) to ensure the evenness and falls of the finished paving. The paving is then 'whacked' down with a compactor and more sand brushed into the joints.

You can have fun with additional buildings or outhouses on your land but make sure they are on the original plans you submit for planning permission.

Paving should be laid to falls, so that the water runs to a drain or a soakaway. Puddles leave unsightly marks and may cause local washing-out of the sand between paving units, loosening of the blocks or slabs, and softening of the subbase. Ideally, rainwater should soak into the ground rather than overload the sewerage system, so if you are building on well-drained soil, consider 'permeable paving'. This does not have to be laid to falls, as the rainwater soaks through it and into the ground.

The integrity of any paving depends on its edgings. The borders of the paved area are laid in a concrete foundation, and can match the general paving or contrast with it. If you don't want a visible edge it is possible to fix the paving at the edges to the foundation with a strong epoxy mortar. Without an edging, paving tends to spread and fall away at the sides, leading to progressive collapse as water gets in and washes out the subbase. On deep, weak soil, another approach is to cast concrete slabs onto which the paving is fixed. These need joints to avoid cracking if any movement occurs.

OTHER GARDEN BUILDINGS

Gazebos, bowers and the like can take the concept of outdoor 'rooms' a stage further. There are off-the-peg models available from builders' merchants and garden centres, but a bespoke design will be more satisfying and is more likely to blend in to your particular garden. You might even consider a fake ruin or 'folly' as a talking point.

WILL I NEED PLANNING PERMISSION?

Boundary fences and gates, referred to by the Planning Department as 'means of enclosure', usually need to be confirmed as part of the planning consent for a house. Anything less than 1 m high facing a highway or 2 m high elsewhere is 'permitted development' (see page 49). This excludes extending in height existing walls and anything that could block views on a highway. If your site entrance crosses a pavement, the norm is to maintain a clear 2400 x 2400 mm triangular 'vision splay' (an area clear of visual obstructions) above 1m from the ground, where the driveway meets the pavement. This is to avoid collisions with people walking on the pavement. In rural areas, where driveways tend to issue straight on to highways, larger vision splays may be required. If your site is big enough, or the highway busy enough, the Planning Department may insist that you provide space to turn vehicles round so that they do not have to reverse on to the highway.

Permitted development rights can be withdrawn by the Planning Department, and do not apply in conservation areas, where any means of enclosure needs consent. This is also true for the surroundings of a listed building, where paving is also controlled. The obvious and safe approach is to check with the Planning Department.

WILL I NEED TO COMPLY WITH BUILDING REGULATIONS?

Roofless structures, such as walls and fences, do not normally fall under Building Regulations unless they present a threat to health and safety. Paving may be affected, however, requiring level, or only slightly sloping, paving from the site entrance to a door into the house. The car-parking area will have to be wide enough for a car and a wheelchair. Paving is required to be 'firm and even'. Steep slopes may be exempt from 'level' access, in which case the design of paths and steps will be controlled to make sure that they are wide enough, not too steep, and that the ground is even; handrails may be required. (Building Regulations cater for people with all disabilities, and with the effects of ageing, and not just with wheelchair mobility.)

Garages

A garage can have many uses other than providing somewhere to park the car: it offers a great opportunity for extra storage or work space, can easily be converted into a living or playing area at a later date, and presents a convenient outbuilding in which to carry out unacceptably messy maintenance work.

DO I NEED A GARAGE?

There are arguments both for and against storing a car in a garage. On the one hand, in a coastal location, it is better for a car to have some shelter, as salt air is very corrosive. Similarly, exposure to the

sun can fade upholstery and perish seals, and droppings from birds and trees can damage paintwork. A garage can also provide a secure place to keep your car and garden machinery. On the other hand, parking a wet (or salty) car in a warm, poorly ventilated environment can accelerate corrosion.

On balance, the best environment in which to store a car is probably a carport. This can be cheap to erect as well, although it must be strongly built to resist wind damage, and relatively unobtrusive. On the negative side, a car port has few of the advantages of a garage, such as conversion potential, providing storage space or security.

WILL A GARAGE ENHANCE THE VALUE OF MY PROPERTY?

A garage, particularly a double one, is a desirable feature, but the cost of building one tends to be similar to the value it adds. A good strategy is to include one in your planning permission, even if you do not intend to build one, as that in itself will add some value to the property. However, be aware that some planning departments may give planning permission with conditions attached – for example, that it can only be used as a garage or for storage and cannot be converted into living space at a later date.

CAN I THINK ABOUT IT LATER?

A garage needs to be thought about at a very early stage of the planning process because it will have to have planning permission, Building Control approval, and will need to be completed along with the house if you want to be exempt from paying VAT (see page 19).

WHERE SHOULD I PUT IT?

A garage can be erected as a separate building, attached to the house, or integrated into the ground-floor plan.
- **Detached** A detached garage doesn't spoil the design of a house, which is especially important with historical styles in which garage doors might look out of place. Visually, a detached garage can complement a house, looking like a farm outbuilding and even creating a sense of a courtyard.
- **Attached** The main advantage of attaching a garage to the house is its potential for conversion

into an extension of the living space if planning permits. It is also cheaper and allows you to get to and from your car under cover.

- **Integral** The integral option has the same advantages as the attached garage, but can look odd in a small house – as if the car has been given a room of its own at the expense of the occupants' living space.

WHAT SORT OF DESIGN SHOULD IT BE?

The safest design solution is to match the house in terms of finishes, roof pitch, joinery style and colour. There are cheaper options, such as erecting a timber or concrete prefabricated garage, but these are difficult or even impossible to convert, and do not add the value of more permanent structures. Most garage doors are made of pressed metal and are rather utilitarian in appearance, although wooden doors are available, or metal-framed ones that can be clad with timber. It is fairly standard practice to put windows and/or a side door in a garage. Roof lights actually give better light, but if the garage is converted, you will need to add windows in the long run.

HOW MUCH SPACE DO I NEED?

The standard parking space of 2.4 x 4.8 m is inadequate for many larger cars, and standard garage door openings are too low for some taller vehicles. If you want to walk around the vehicle or add a workbench (and use it when the car is inside) you should add at least 1.5 m to the inside length or width. Bicycles tend not to be thought about at the design stage, and end up being wedged in next to the car. You may want to consider dedicating some space for them.

Take into consideration the amount of space you need in front of the garage in order to drive your car in and out safely. The narrower the doors, the wider the turning space has to be – 6 m is a bare minimum if you are turning 90 degrees into the garage. On busy roads you will probably be obliged by the Planning Department to allow space in front of the garage in which to turn your car round, so that you can both enter and leave your property driving forwards.

HOW SHOULD THE GARAGE BE BUILT?

A garage is a cheaper space to build than a house, because there are no requirements for insulation.

However, if you want to build in the potential to convert, and the Planning Department will allow you to do so, you should use a structural system that can be insulated later. Most structures can be lined with thermal laminate, a lining board bonded to a layer of insulation. Think also about how you might add a damp-proof membrane to the floor. For example, using a wide damp-proof course in the walls, with 150 mm left exposed, would allow you to bond a damp-proof membrane to it later. If the membrane is 150 mm up from floor level, you will then have enough space to insulate the floor. Make sure you allow enough headroom, and place doors and windows so that they will be at a usable height if the floor level is ever raised by the addition of insulation.

If you intend to convert space in the pitched roof, your walls will have to be strong and stable enough to support bigger loads, and to ensure this you will need the advice of a structural engineer. Otherwise, you can use fairly thin walls. The Building Regulations Approved Document 'A' contains advice on the thickness of walls. The longer or taller your garage, the thicker the walls have to be in order to be stable, but you can add buttresses for more support.

To allow for future conversion of the roof, you will have to design the garage so that it is high or steep enough to hold usable space. Make an allowance for the considerable thickness of insulation you will need at the same time. Think about how the roof will be ventilated if you do intend to add insulation later, and whether the rafters are big enough to support such insulation. Without such future-proofing measures you may end up having to rebuild the whole roof in the event of a conversion.

Make sure that the foundations are adequate – prefabricated or lightweight garages require minimal foundations, which are not suitable for conversion. In addition, bear in mind that it is cheaper to provide drainage and water and power supplies at the time of building than when you come to convert it later on.

Design checklist

GETTING THE STYLE AND LAYOUT RIGHT

☐ Check to see what kinds of house design your local authority will allow.

☐ Choose a style that suits your personal taste.

☐ Decide on the precise size, shape and position of the house on your plot.

☐ Work out the layout of each room on graph paper.

☐ Decide on a theme for all your fixtures and fittings to give your design cohesion.

☐ Look for refurbished fittings where they might give your design a more authentic feel and save you some money.

☐ Put the same thought and planning into your landscaping as you do your internal design.

USING A PROFESSIONAL

☐ Hire a professional designer or get professional advice on your design early on to make sure you meet town planners' requirements.

☐ Establish your and your professional's responsibilities and price at the outset.

☐ Remember it is your house: your designer should work with your ideas not impose his/her own.

BUREAUCRACY

☐ Draw up a site plan to show how your house sits on your site and how your site relates to surrounding properties and features. Make sure your plans comply with Building Regulations.

☐ Inside, double-check that your fixtures and fittings comply with local planning regulations, particularly if you are renovating a property that is listed.

☐ Outside, check that your boundary fences and gates meet planning requirements and that these are confirmed in the planning consent for your house.

☐ Consider incorporating your hard landscaping design, including conservatory, garage and other external buildings, in your initial build to avoid paying VAT.

DRAWING UP PLANS

☐ Draw up a ground-floor plan to show the layout of your house plus adjacent landscape features.

☐ Draw up a first-floor plan to show the room layout, plus some structural information such as joist directions and beams in floors.

☐ Draw up a front-elevation plan to show the heights of your ground and first floors and more detailed dimensions, including your soil vent pipe location.

4 Building Regulations

If you are embarking on any kind of major building work on a house – whether a new build, renovation or conversion – you will come across Building Regulations. These are legal requirements aimed at ensuring that all building and construction works achieve the required minimum standards. Current standards are laid down by the Government in the Building Act of 1984 and in regulations made under the Act, including the Building Regulations 2000. Building Regulations are distinct from planning control, which is concerned with the use of the land, the appearance of the building and its effect on the neighbourhood and environment (as discussed in Chapter 2).

Understanding Building Regulations

The Building Regulations 2000 define the types of building work that are subject to the regulations and those that are exempt. They set out procedures to be followed when undertaking building works and specific requirements for aspects of building design, such as the health and safety of building users. Building Regulations do get modified, so it is best to check the latest information with the Building Control Department at your local district council.

Practical guidance on how to meet the requirements of the Building Regulations is contained within a series of documents known as Approved Documents, each of which relates to a specific part of the regulations. Two recent Approved Documents are Part L and Part M. Both of these have had a significant impact on house design since their introduction, because they deal specifically with energy conservation and access for the disabled (see pages 96–101).

Building Regulations have three main purposes:
1 to ensure the health and safety of people in and around the buildings
2 the conservation of energy
3 access and facilities for disabled people.

If you are building a new house, or carrying out major works to an existing one, you cannot avoid Building Regulations: your mortgage provider will need a copy of the completion certificate issued by your building control officer and, if you choose to sell your property, your buyer's solicitor will certainly require a copy. If you cannot provide it, you can expect the sale to fall through or to have to negotiate on the purchase price.

WHEN DO I NEED BUILDING REGULATIONS APPROVAL?

Building Regulations approval is required for all new building works (including extensions) and refurbishment/renovation work (but not repairs). Some works are exempt and these are listed in Schedule 2 of the regulations. They include the following:
• A new single-storey building built of non-combustible material, with a floor area of less than

All new-build developments, conversions and extensions have to comply with Building Regulations. These set out specific requirements regarding the design of the building.

30 m^2, no sleeping accommodation, and not less than 1 m from the boundary of the site. Extensions to existing properties are not covered by this exclusion.

- A new detached building with a floor area of less than 15 m^2, which has no sleeping accommodation.
- Ground-floor extensions, such as a conservatory or porch, as long as the floor area is less than 30 m^2 and as long as any glazing meets the requirements of Part N of the Building Regulations.
- Non-structural alterations that do not result in any changes to the means of escape or access for the disabled. For example, if you are relocating or replacing the staircase you will need approval, but if you are splitting a large bedroom into two with a simple partition you do not.

If you have any doubts as to whether you will need approval for your project, it is best to contact a building control officer at your local district council. Explain your project and ask for his or her advice.

HOW DO I GET BUILDING REGULATIONS APPROVAL?

Building Regulations and the approval of projects fall to the Building Control Department of your local district council. It is their role to review applications for compliance, to issue advice, to visit the site at regular intervals to check adherence to the regulations and to issue the final completion certificate. 'Approved inspectors' can also fulfil this role: these are private companies that that have the authorization to grant Building Regulations approval in the same way as your local council.

Although Building Regulations are a statutory piece of legislation, each local authority may have slightly different procedures, although their interpretation of the regulations should be the same. There are two methods of applying for Building Regulations approval: a building notice or a full-plans application.

In the case of a building notice, you do not submit any drawings or documents but notify the local authority that you intend to commence a project. The building control officer then approves the work as it progresses. This may sound simple, but should only be considered by experienced builders. Because you are not working to approved drawings, you only get approval when the officer comes to the site and inspects the work. This means that the pace at which you can work is governed by the frequency of the officer's visits. Should you progress beyond a stage that has not had approval, you are taking a considerable risk.

The full-plans route involves you submitting your working drawings and specifications to the Building Control Department, together with a completed application form and payment of the specified fee. Depending on the information shown on the drawings, you may also need to provide further information – calculations from your structural engineer, for example. A building control officer will be assigned to your application (he/she will also be the officer who inspects your build). This officer will then review your application and drawings and check that the information provided demonstrates that your proposed work meets Building Regulations requirements.

Following this review, the officer will write to you and advise you whether your application has been accepted or rejected. When an application is rejected a letter is sent, listing the areas where the design has failed. You can speak to your designer and ask him or her to address the reasons for the rejection. Once they have been corrected you can resubmit your drawings and the case officer will review them again.

WHEN DO I APPLY FOR BUILDING REGULATIONS APPROVAL?

It is important to wait until you have detailed planning permission for your build before you submit for Building Regulations approval. The two applications are linked, and your Building Regulations application drawings need to reflect the project drawings, for which planning permission has been granted.

Most local authorities aim to complete the review procedure within 4–6 weeks of an application being submitted and formally registered, and it could be as soon as 2–3 weeks. When calculating your schedule, it is recommended that you allow an additional 2–3 weeks should your application be rejected and design alterations are necessary in order to satisfy the officer.

It is highly recommended and desirable to have full approval before you commence any work on site. If you have to carry out any demolition, this can be done before you have approval, but do not commence excavating foundations.

It is possible to add a conservatory to a house without
Building Regulations approval as long as it is no larger
than 30 m² and meets other restrictions.

HOW MUCH WILL IT COST?

There is no set scale of charges for Building Regulations applications. Ask the Building Control Department to send you a schedule of charges. Most authorities' fee charges are based on the gross floor area of the build (that is the measurement from the external walls), and consist of two separate fees: an application fee to review your documentation and an inspection fee, which covers all on-site inspections during your build. These fees are not high: for a 186 m^2 house expect to pay around £150 for the application fee and a similar amount for the inspection fee.

WHAT DO I NEED TO KNOW?

Many projects fall foul of Building Regulations because they are under constant revision, and tradesmen often find themselves working to out-of-date regulations. It is also important to know that, once your drawings have been approved and work has commenced on site, the building inspector cannot make you carry out works that are not detailed on your drawings. If a mistake is found on your drawing or something has been missed out he or she cannot force you to rectify it during the build, although he or she may request that you do so.

WHAT HAPPENS IF I BUILD WITHOUT BUILDING REGULATIONS APPROVAL?

Your Building Control Department has to see that all building work complies with the regulations. If you proceed and carry out work without approval the department may ask you to alter the work that you have done, or even remove it, at your own expense. If you fail to do so, they may serve a notice on you. There is an appeal process, but it is best not to get into this situation in the first place.

The inspection process

Depending on your local authority, the building control officer may also represent the Planning Department. During each visit he or she will check that you are building to the approved drawings in respect of both Building Regulations and also your planning permission.

If you are carrying out a renovation project the inspections are more flexible and it is recommended that you keep in regular contact with your officer.

Let him or her know your schedule and he/she will visit when necessary. The stages at which building work is inspected by the building control officer are dependent on the type of building work being carried. Usually, on a new build, the following will be the minimum visits that you can expect:

• commencement
• foundation excavations
• foundation concrete
• concrete ground-floor slab
• damp-proof course
• drainage
• roof completion
• completion.

COMMENCEMENT

You must notify the Building Control Department of your proposed start date. Try to give the officer at least seven days' notice, 14 if possible. Most officers like to visit the site on the day that works commence, and this is something that should be encouraged. During the first visit the main objective is to familiarize him- or herself with the project. Most building control officers are assigned to projects in their region. If you are using a local builder or subcontractors there is a very strong possibility that they have worked together before, which should help on-site relations. The inspector may check any setting out that you have established to ensure that you are positioning your house according to your planning drawings.

A building control officer will inspect your site a number of times during the build to make sure that you comply with Building Regulations at each stage of the process.

EXCAVATION OF FOUNDATIONS

This is one of the most critical visits the building control officer will make, for it is his or her responsibility to make sure that your foundations are deep enough to support the considerable weight of your house. Your groundworkers will use their experience and dig the foundations to a depth that they feel is required, taking into account the ground conditions and the proximity of nearby trees. Before concrete is poured into the foundations, the building control officer needs to approve the depth of them. If you pour concrete into the foundations before getting the building control officer's approval, you could be facing a very expensive bill for either removing the concrete or excavating trial pits around the foundations to prove that they are a correct depth. The officer will be looking for a number of things on this inspection. For example, have the excavations been dug down to solid ground? Is there any evidence of any roots or vegetation in the excavations? The foundations are vital to the structural stability of your house, and the building control officer's recommendations are, therefore, to be taken seriously.

Establishing a good relationship with your building control officer will make site inspections easier for all parties.

CONCRETING OF FOUNDATIONS

The building control officer may choose to pay a further visit when the concrete is being poured, to make sure that it is of the right mix and is being poured correctly. It is possible for the mix to 'segregate' during pouring, which means that the cement aggregates and water that form the concrete separate, and this can have dramatic effects on its strength and long-term characteristics, resulting in serious problems with the performance of the foundations and their ability to support the structure of the house.

CONCRETE GROUND-FLOOR SLAB

The officer will be keen to establish that, prior to pouring the ground-floor slab, you have cleared the oversite (the area within the perimeter of the walls) of any vegetation and that your damp-proof membrane has been correctly installed to prevent unwanted transfer of moisture from the ground into the concrete slab.

DAMP-PROOF COURSE

On this visit, the officer checks that your bricklayers have installed the damp-proof course correctly. He or she needs to check that it is at the correct level, is not ripped and is continuous along the perimeter of your house.

UNDERGROUND DRAINAGE

The building control officer will check that all the drainage has been installed in accordance with Building Regulations.

The main issues here are: that it is laid to a fall to allow the flow of waste away from the house; that inspection chambers are installed at any bends (to allow for cleaning and rodding should a blockage occur); and that the joints between the pipes have been made correctly. The officer may ask for an air leakage test to be carried out. This is where the drain runs are sealed with bungs and a simple pressure gauge is applied. If the pressure drops there is a leak, and the test has failed. It is advisable to have this inspection made before the drainage is backfilled. Should there be a problem not only will there be a delay, but you will have the additional expense of removing the backfill to enable you to correct the problem.

The building control officer's visits are fundamental to the success of your build and guarantee that things such as drainage are completed to a sufficient standard.

ROOF COMPLETION

The officer may choose to make a further inspection when the roof has been completed. This is to check that the roof trusses have been correctly braced and have been securely fixed to the blockwork skin.

COMPLETION

Once your house has reached the stage where you consider it complete, your officer will need to carry out a completion inspection of the site. During this inspection he or she will carry out a full check, including the windows (for fire escape purposes), glazing (for safety reasons), roof-space insulation, smoke detectors, ventilation and various drainage tests. Should everything be in order and the officer is happy that the finished house is in accordance with your approved drawings and the regulations, then he or she will issue a completion certificate.

Approved Document Part L: conservation of fuel and power in dwellings

There is growing concern that, with high usage rates, the world's energy resources will become significantly depleted. It has been recognized that our homes and buildings are large users of energy (oil, water and gas) and that architects and developers have an obligation to design buildings in such a way as to minimize their use.

Part L has five main points:

- Provision must be made for the conservation of fuel and power in a building.
- The fabric of the building must limit heat loss.
- Controls must be installed to operate the heating of the building and hot water.
- Heat loss from hot-water tanks and central heating should be limited.
- Lighting systems should not use more energy than needed and should have adequate controls.

This approval document also applies to renovation projects where you are carrying out substantial works in replacing major elements of a roof, floor or exposed walls. These points mean that more insulation is required, windows and doors must be energy-rated (this even applies to new windows in existing houses), and all boilers (in both new and existing houses) must be energy-rated (see below). Although these points may sound daunting, there are long-term benefits for you in your new home. You will pay more money at the time of building, but in the long run you will use less energy and will have much lower energy bills. Furthermore, it is actually relatively easy to achieve these new requirements because manufacturers have been very quick in bringing out new products that meet the levels required. (Look for products that are advertised as being 'part L compliant'.)

Heat loss to buildings is measured using a term called the U-value, which calculates heat loss in watts per metre squared of a building (walls, floor or roof). These U-values are a measure of how much heat will pass through 1 m^2 of matter when the air temperatures on either side differ by $1°C$. The figure is expressed as watts per square metre per degree centigrade. This is abbreviated to W/m^2K. If, for example, an element has a U-value of 1.0 (written as $1.0W/m^2K$) that element loses one watt of energy for one degree centigrade difference between the inside and outside. The approved document sets out the maximum U-value that has to be achieved – the lower the number the lower the heat loss.

HOW DO I ACHIEVE THESE VALUES?

A competent house designer should design your house to meet these levels, and many will exceed them. However, what should you look out for? Below are some of the most common ways of achieving the right insulation level depending on your type of build.

- **Brick cavity walls** Historically, the cavity between the brick outer skin and the block inner skin was left empty in the belief that a 90–100 mm air-filled cavity was a good insulator. By the 1980s

CHANGES IN U-VALUE REQUIREMENTS

Building element	Previous U-value W/m²K	2002 U-value W/m²K resistance	% increase in thermal performance
Walls	0.45	0.35	28.6%
Ground floor	0.45	0.25	80.0%
Flat roof	0.35	0.25	40.0%
Cold pitch roof (loft)	0.25	0.16	56.0%

and 1990s the cavities tended to be pumped with foam. More recently, as better insulation is required, it is common to put 90 mm of insulation in the cavity (normally an insulation quilt or foam panels) as the wall is constructed.

- **Timber-frame houses** It is the norm to fill the voids between the timber frame with insulation. To achieve the required U-value levels, this means having 140 mm thick studs if you are using mineral wool, such as glass fibre. Alternatively, you can use a 90 mm frame with high-performance insulation boards or add more insulation to the internal side of the building.
- **Roof insulation** Historically, 150 mm of glass-fibre insulation (or similar) was laid between the joists. Now you need to lay an additional 100 mm over the joists in a criss-cross pattern.
- **Windows and doors** All new windows and doors installed in new houses or as replacement windows now need to meet the requirements.
- **Boilers** All boilers now have to meet the new requirements. Speak to your plumber and make sure that you install the most efficient boiler available to you.

WATER CONSERVATION

The approval documents include water as energy and have attempted to reduce the amount of water used in our homes. This is done mainly through the adoption of low-flush and dual-flush toilets.

Open areas may be prone to greater heat loss but improvements can be made through the effective use of better insulation and draught-free windows.

THE SAP RATING

You need to provide evidence that your house meets with the requirements of Part L, and this is achieved by providing a SAP rating.

SAP is the Government's Standard Assessment Procedure for energy rating of dwellings. SAP was designed to be included in the 1995 Building Regulations and it is now a compulsory component under Part L. Every new house has to have a SAP rating, which can be defined as follows:

- SAP provides a simple means of reliably estimating the energy-efficiency performance of dwellings. SAP ratings are expressed on a scale of 1 to 120: the higher the number, the better the rating.
- SAP is calculated by a procedure contained in the Building Regulations, which predicts heating and hot water costs. These depend on the insulation and air tightness of the house and the efficiency and control of the heating system.
- The procedure for calculating SAP is defined by the published SAP worksheet – though in practice most people calculating SAP use one of the Building Research Establishment (BRE)-approved SAP calculation computer programs.
- SAP programs are used to enter data on the size of the house, its insulation levels, its ventilation system and its heating and hot-water systems. SAP rating can then be submitted for Building Regulations approval and is checked by the local Building Control Department.

Calculating the SAP rating should be carried out by your designer or package company. You may need to provide additional information, such as the boiler specification (from your plumber), to enable him or her to correctly carry out the calculations.

You also need to demonstrate that your house has been designed to meet the specified maximum U-values. This is calculated using three methods – all of which are submitted to the Building Control Department. It is quite common to fail two of the methods and pass on just one. This is quite acceptable and will not result in an unsuccessful application. The three methods are:

Method 1: Elemental method

This is the most straightforward method for compliance, as it relates to the specific U-value

---

REQUIREMENTS OF THE 2002 APPROVED DOCUMENT PART L

2002 minimum U-value requirements

England and Wales

Floors	0.25
Walls	0.35
Pitched roof insulation between joists	0.16
Pitched roof insulation between rafters	0.20
Flat roof	0.25

Scotland

Floors	0.25
Walls	0.30
Pitched roof insulation between joists	0.16
Pitched roof insulation between rafters	0.20
Flat roof	0.25

Northern Ireland

Floors	0.45
Walls	0.45
Pitched roof insulation between joists	0.25
Pitched roof insulation between rafters	0.25
Flat roof	0.35

Republic of Ireland

Floors	0.25
Walls	0.27
Pitched roof insulation between joists	0.16
Pitched roof insulation between rafters	0.20
Flat roof	0.22

---

requirements for each element of the building (see table above showing the 2002 minimum U-value requirements). Therefore it compares, for example, the required U-value for external walls – say 0.35 – with the actual U-value for the external walls of your house.

Method 2: Target U-value standards

The target U-value (μ) method of compliance addresses the thermal performance of a building as a whole. It provides the house builder with a more flexible approach than the elemental method and can be used with any heating system. In considering the overall U-value for the dwelling, this method

Rooms with glass walls and ceilings have low SAP ratings because they suffer from heat loss. Other rooms may score highly, however, and counterbalance such low scores.

A vertical chrome radiator like this will not only heat your bathroom, but will also provide somewhere to store and dry your towels.

HOT-WATER AND HEATING SYSTEMS CERTIFICATE

Installation of all the hot water and heating systems in your house need to be fully commissioned. This means that they need to be set up in full working order and need to be tested to demonstrate this. Following this commissioning your plumber needs to issue to you a certificate confirming the correct installation and commissioning – this can be the Benchmark Code of Practice for the Installation or Commissioning and Servicing of Central Heating Systems. You as the home owner must also be given full information and instruction on the operation and maintenance of your hot water and heating systems.

allows for trade-offs between openings of windows, doors, roof lights and other insulation levels. Therefore if you have a glass elevation (say for example in a barn conversion) where the glass provides an unsuccessful result using the elemental method, you can use the target method to trade off this area of lower performance with areas of higher performance to achieve the required level.

Method 3: Carbon index

The carbon index method provides the greatest flexibility of the three. It is based on the annual carbon-dioxide emissions associated with space and water heating. This will be influenced by your heating systems and you need to be sure that they have been correctly specified and installed.

The carbon index is based on the total annual CO_2 emissions associated with space and water heating per square metre of floor area. It is expressed as a number between 0.0 and 10.0 rounded to one decimal place. If the result of the calculation is less than 0 the rating should be quoted as 0; if it is greater than 10, the rating should be quoted as 10. To comply with Part L, the dwelling must achieve a carbon index of at least 8.0.

The carbon index is linked to your SAP rating and the calculation involves the use of computer software programs, therefore it is recommended that this calculation, if this route is chosen, is carried out by the person who calculates your SAP rating for building control.

A whole article could be written about SAP ratings and how they are calculated but it should not be a major concern as long as your house has been designed to reflect the new Approved Document L.

Approved Document Part M: Access and facilities for people with disabilities

The intention of Part M is to ensure that reasonable provision is made for disabled people to gain access to, and make use of, a building. Your design needs to meet all the requirements listed. Key areas are:

- **Disabled access** A change in level at ground floor, internally or externally, needs to be made with a ramp and not steps. There must be a solid path from your site boundary to your front door to allow wheelchair access to your home. A ramp to the front door must be 900 mm wide and 1.2 m long to allow a wheelchair to manoeuvre.

- **Door opening sizes** All ground-floor and upper-floor door-opening sizes need to have a clear opening of between 750 mm and 800 mm. The front door has to have an opening of at least 775 mm.
- **Downstairs cloakroom** All new homes must have a downstairs toilet. You need to provide a room in which a wheelchair can enter and turn: this means that the door has to open outwards. Current regulations contain the minimum dimensions for the size of this room.
- **Electrical sockets and light switches** These now have to be positioned so that they can be reached by someone in a wheelchair. A band from 450 mm to 1200 mm above finished floor level is specified for the installation of switches and sockets. (Light switches, doorbells and entry phones at a maximum of 1200 mm; electrical sockets, TV sockets and telephone points at a minimum of 450 mm.)

HOW DO BUILDING REGULATIONS DIFFER IN SCOTLAND AND NORTHERN IRELAND?

Despite being part of the United Kingdom there are differences in the Building Regulations from England and Wales to those in Scotland and Northern Ireland. In many instances these tend to be cosmetic, for example, Approved Document Part L in England and Wales is called Part F in Northern Ireland.

However the Scottish regulations call for a lower U-value than in England and Wales, while, in Northern Ireland, a higher U-value is required. Building Regulations now require a U-value for walls of 0.35 W/m²K, 0.16 W/m²K for lofts, and 0.25 W/m²K for floors (England and Wales). For walls the requirement in Scotland is 0.30 W/m²K; and in Northern Ireland 0.45 W/m²K.

WHERE CAN I GO TO GET HELP?

The Building Regulations and the associated Approved Documents can seem daunting and confusing. If you are employing an architect or technician to design your house you should make it his or her responsibility to ensure that your house is designed to meet all the requirements of the regulations. If you are presented with any problems or have any queries, simply relay them back to your designer. You can also seek help and advice from the Building Control Department of your local authority. A third route is to try materials manufacturers: if you need information on how to use materials to meet the Building Regulations, most manufacturers have a technical support department and will almost certainly be able to provide you with the advice and technical information needed to find a solution if you have a problem.

All new builds have to provide disabled access to the house. This means that a ramp has to be installed wherever there is a change in level on the ground floor.

5 Construction methods

When it comes to building your house, you have certain choices over the type of construction you use. To an extent, this will be dictated by the design of the house or the site, but you, your house designer or other building professional will also influence this decision.

The basic construction methods

This section offers an overview of the main types of shell construction – brick and block and timber frame – and offers more detail on the various aspects of building a house: the foundations, walls, floors, roofs and insulation. The intention is not for you to construct the house yourself, as most elements of the build should be done by a professional. Instead, the professionals you work with will be able to advise you on the finer details, and the information here will be a useful reference for the terms your architect, structural engineer, builder or other professional might refer to at any give stage of the build.

Choosing what you would like to build your house from doesn't have to be complicated. If you choose a house from a design and package company, for example, the provider will help you decide on the best materials to use. If opting for a more independent approach you may want to consider choosing the construction method yourself, in which case it is important to decide early on if you have a preference, because this will affect your design, your budget and your build time.

There is more than one way of building a domestic house, but the majority of houses are built using either a brick-and-block or timber-frame construction. Both of these techniques benefit from the latest advances in technology and development and are reliable methods of construction that have been in use for hundreds of years.

Although both methods are used throughout the UK, brick and block is still the method most favoured by house builders in England, Ireland and Wales, while timber-frame is the most popular in Scotland. Choosing between the two methods used

Your choice of construction method will be influenced by all manner of criteria, not least budget and timing. Use professionals to help you come to the right decision.

to be a lot simpler but, today, either can be used for most new builds, unless you have an unusual design, in which case the design will dictate the construction method.

BUILDING WITH BRICK AND BLOCK

Brick-and-block (or masonry) construction involves building from the ground up, starting with the foundations and finishing with the roof. Its name derives from the fact that the house is constructed with two walls, the external one made from bricks and the internal one from blocks. The internal and external walls are separated by a cavity which is usually filled with insulation. The brick-and-block walls rise from the foundations for two or three rows of bricks before the damp-proof course (DPC) is added. The DPC is a water-resistant membrane that sits between those first few layers of brick-and-block walls and the rest of the house rising up from it.

Typically in a brick-and-block construction the inner skin is concrete block, while the outer skin may be brickwork, rendered block or stone.

The inner wall, built from blocks, shoulders the majority of the weight of the building and keeps moisture out and heat in. The blocks need to be light, strong and provide thermal and sound insulation to meet Building Regulations. They are available in a variety of shapes and sizes suitable for external and partitioning walls, floors and foundations, and are made either from aggregate concrete or Aircrete (lighter bricks made of a compound that includes cement).

The external wall provides weather resistance and the look of your house and the most common choice is brick although, subject to planning permission, it could also be built using stone or clad with timber weatherboards or tiles.

The properties of this structure provide good thermal and noise insulation and this may be complemented by the addition of sound-insulating panels. Brick-and-block construction also provides good fire resistance.

New building techniques, such as thin-joint mortaring, have helped speed up the brick-and-block building process, while a design sympathetic to this method will help keep the job simpler and quicker – having wall dimensions that won't require blocks and bricks to be cut, for example.

What are the advantages of brick and block?

The majority of UK builders are most familiar with this form of construction, so finding skilled professionals to do the work won't be a problem. If you want an unusual shaped house, for example one with curved walls, brick and block is a good choice. It is more flexible if your house is built on an uneven plot or if there are obstructions that need to built around (trees or posts). In fact any unevenness is easily remedied on site as the construction goes up. Brick-and-block construction tends to be cheaper, unless the house design is very complicated. At the very least, the cost of the construction is spread more evenly throughout the build.

What are the disadvantages of brick and block?

Brick-and-block houses tend not to support their weight entirely in the external walls, meaning some of the internal walls need to bear it too. This places some constraints on just how much open space you can have inside a brick-and-block design. If you are keen to have a large open-plan area, for example, you will probably have to incorporate steel beams to spread the load to the outer walls. Alternatively, an increasingly common practice is to have concrete floors above ground level, which allows the floors to spread the load to the external walls and reduces the need for supporting internal walls. During the building work, your house is vulnerable to the weather until the closing stages, say 18–24 weeks.

BUILDING WITH A TIMBER-FRAME

With this method, a house is built around a timber-frame. The frame of the external walls supports the whole structure, so there is no need for internal supporting walls. The frame is prefabricated at a factory along with any number of other components, including wall and floor panels. With both off-the-shelf and bespoke designs, it is possible to have the house delivered to your plot as a quick-build kit that can be roofed and weatherproof in just a few days.

You can dress a timber-frame house as you wish and it is quite possible to put a brick fascia on it to conceal the timber entirely. Most often, however, the timber-frame is a feature of the design. Douglas fir and sitka spruce are the mostly commonly used timbers and come from managed renewable forests in North America, Scandanavia and southern Germany. Top-of-the-range timber-frames are built from oak and you can expect to pay rather more for your wood.

While a timber-frame might not have the bulk of a brick-and-block construction, it does have the same strength and insulating properties. With age the timber contracts, tightening the joints and strengthening the whole frame, and wood holds heat extremely well. Furthermore, all manner of panels and materials can be added to improve heat and sound insulation, fire and water resistance, and to add strength. As with brick-and-block constructions, all timber-frame building materials have to meet the necessary Building Regulations before assembly starts.

What are the advantages of timber-frame?

Timber-frame construction offers the opportunity for a quicker on-site build time than brick and block, although this does rather depend on the design. The building will also reach a watertight position more quickly, which means you are less dependent on the weather to take your build to the next stage. Consequently, there is greater flexibility in the order of building work. Timber-frame construction also offers big open-space opportunities inside your build, and people tend to choose this method if they like to see timber posts and vaulted ceilings. If you choose your timber carefully, that is from a renewable source, there is an environmentally friendly benefit too.

What are the disadvantages of timber-frame?

A timber-frame house has to be engineered to perfection before it goes on site, so there is little room for modification to the design once the frame has been made. It is also less adaptable to obstructions or defects on plots. Fewer skilled professionals are used to this type of build in England, Wales and Ireland, although most design and package companies will help you find professionals with experience. With this option, you will have the up-front cost of prepaying the package company for the frame.

Timber-frame buildings offer the advantage of a quicker build. You can buy an off-the-peg package, which can either be assembled on site or erected as prefabricated panels.

BUILDING WITH POST AND BEAM

A variation of timber-frame construction, the post-and-beam approach was characterized by the homes of the New England colonists of the 17th century and is very popular in America to this day. While modern timber-frame techniques use lighter, softwood beams and metal clips and hangers to butt-link the joists, post and beam favours chunky hardwood beams slotted together and held in place by wooden pins hammered through them. It has a certain old-world charm that lends character to a design.

Like a more conventional timber-frame, a post-and-beam construction carries the weight of the building and allows for wide open spaces inside your home. The big hardwood beams make a lovely design feature and can be exposed both internally and externally. Alternatively, you can completely conceal the post-and-beam frame with internal and external dressing, including a brick fascia. Post-and-beam frames are prefabricated to your chosen design, and partly because more wood is used in this type of construction, it usually costs more than a standard timber-frame.

CONSTRUCTION METHOD CHECKLIST

Once you have considered all of the options, use the following checklist to confirm your choice of construction method:

- Do your design requirements favour one particular construction method? The external look and interior space of your design may depend on your choice of construction method.
- The location and nature of your plot may dictate your construction method. What building materials are available in the area and what is the access to your site like? Similarly, what can your plot support? If access and ground support are limited, you may not be able to bring in cranes for lifting frames.
- What skilled professionals are available to you? Check whether you will be able find the people you will need near your site.
- Would bad weather delays seriously affect your plans? If so, you will need to time your project carefully and consider which construction method would be quickest and leave you least vulnerable to bad weather.

It is important to remember that timber shrinks. This means incorporating complicated movement joints and seals that need checking periodically.

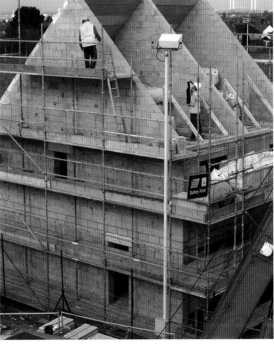

It is now possible to use alternatives to the brick-and-block method – for example the Jamera system being used here – that can be built significantly faster.

MODERN DEVELOPMENTS IN CONSTRUCTION METHODS

Jamera

This system is a variation on the brick-and-block method that does away with the bricks. Jamera uses Aircrete blocks, reinforced with non-stressed steel, and a tongue-and-groove arrangement for aligning and securing them to one another. The blocks are available for internal and external walls, floors and roofs and can span up to 6 m. They come with smooth or textured finishes, or ready for cladding, and are equipped with access for services.

It is possible to buy larger Aircrete blocks, designed for thin-joint assembly. The block size and thin-joint technique makes for fewer, thinner joints, which means improved thermal and acoustic insulation and reduces the need for additional insulating material.

The blocks are assembled with a purpose-made mortar, which is designed to set to full strength within two hours, allowing walls to be built much more quickly than with conventional brick and block. A single block wall may be raised to single-storey height in a day.

These blocks can be used for the block wall in a conventional cavity-wall construction, and can be combined with regular brick-and-block builds as suits. To make the most of the advantages offered by the larger blocks you need to use them in the single-wall complete house build Jamera system. This combines the speed and ease of frame building with the advantages of brick-and-block materials. The building materials are more expensive than regular brick and block but savings are made on labour costs. A typical four-bedroom house can be completed four weeks earlier using the Jamera system rather than conventional brick and block. The system has been developed and used in Scandanavia for many years and is now being used in the UK.

Panablok

Panablok is a technique that takes the frame approach to brick and block a step further. Panablok panels are composed of cement particle boards, which sandwich a rigid urethane core. The tongue-and-groove panels have integral closure mechanisms embedded in the urethane core, to bind them to their neighbours. This system reduces the laying of bricks and the construction process is not affected by weather, so is a good deal quicker. However, the panels need to be prefabricated and do not allow the same freedom of design as conventional brick and block.

Structural insulated panels

Structural insulated panels (SIPs) are made of oriented strand boards sandwiching a rigid urethane core. The panels are dressed internally with vapour-controlling membranes and plasterboard, and externally with brick or timber as desired. SIPs can be used for walls, floors and roofs on a timber-frame. Their large size gives them thermal and sound insulation values that achieve Building Regulation standards without the need for additional panels. Indeed, SIPs can be made with integral joints, which produce a very snug fit and result in air-leakage rates comfortably exceeding Building Regulations. The Kingspan TEK system uses a unique jointing system and a house can be constructed entirely from its

The Jamera system.

Kingspan Tek system's structural insulated panels.

SIPs – walls, floors and roof – without the need for a timber-frame. The insulation efficiency in the TEK system can mean that some houses do not need conventional central heating. Additionally, the TEK system SIPs save on internal space, giving up to 10 per cent more internal floor space on the same plot than a conventional timber-frame build would. With or without timber-frames, SIPs boast very quick assembly although, of course, they need to be prefabricated. Originally, SIPs were practical only for large-scale housing developments but a new flexible modular system now makes them a viable option for individual house builds.

- What demands will your construction method make on your cash flow? If you do not have your budget available up front then you may need to look at cheaper methods that you can pay for over time.
- Whichever construction method you choose, make sure your plans synchronize so that you have the right materials, people and equipment in the right place at the right time.

Foundations and services

A golden rule of building is to find out as much as you can about the ground on which you want to build. There is always uncertainty as to what lies beneath the surface, and dealing with any surprises it presents can seriously upset a budget. There are many aspects to a full and thorough site investigation:

- **Desktop study** Gather evidence that already exists in the form of geological surveys, old maps and records showing previous use, and any existing information from land and property search specialists the Landmark Information Group, your local authority Environmental Health Register of Contaminated Sites and the British Geological Survey.
- **Local knowledge** Consult local people, the local authority building inspector, or anyone else who might know the history of the site, or adjacent sites, and the likely conditions below the surface.
- **Evidence on the ground** Look for telltale signs such as cracks in nearby buildings, hints of flooding, flat areas that might be evidence of old stream beds, rubble indicating dumping or demolition of previous buildings on the site, and recent developments, such as nearby excavations and tree felling, that may affect the moisture content of the soil. Look for indications of buried services – scour the area inside and outside the site for manholes, and if possible, open them up to figure out how they are connected. Little concrete posts with plaques on them indicate buried mains and cables. Ask the local water, electricity, telecommunications and gas infrastructure companies for maps showing the position of their equipment.
- **Ground investigation** To carry out a ground investigation it's best to consult a professional. He or she will probably dig trial pits a few metres

The right kind of foundations for your house will be determined by the 'make-up' of the ground on which you are building.

Foundations will support the walls and roof of your home and provide a solid base on which to build, so it is important to choose the right kind of foundations.

deep in order to assess what the soil quality is way below the proposed foundations of your new building.

- **Boreholes** These are holes that are drilled and cores of soil removed and analyzed. They can go deeper than trial pits, or can be started from the bottom of a trial pit.
- **Soil analysis** Samples from trial pits or boreholes can be analysed for such things as moisture content, strength and the presence of toxins or chemicals that might damage the foundations.

Where should I start?

This can present you with a dilemma, in which the ground conditions dictate the foundation design, while the likely foundation design dictates the kind of ground investigation (particularly the depth, but also the tests to be carried out on the samples). The likely depth of foundations is dependent on how much water there is in the soil, on weak or unstable upper layers and on trees – all of which may be apparent without having to dig. You may ask the advice of the site investigators themselves, or enlist the services of a professional engineer. There will always be an element of risk. There may be sufficient evidence gathered above the ground to give you enough confidence to dig foundations and merely have the local authority building inspector look at the trenches and confirm the assumptions. Trial pits and boreholes reduce the uncertainty a little more, but cost money: trial holes might cost a few hundred pounds, a

borehole and full analysis over £1,000. Furthermore, they only reveal the conditions at their own positions and not across the whole site, although soil analysis can turn up evidence of contamination before expensive mistakes are made or unrealistic budgets set. Specialist companies, listed in directories under site investigation, land surveyors or geotechnical surveyors, carry them out.

THE RIGHT FOUNDATIONS

The ideal ground on which to build a house is rock. If this is not available, then the house will be built on soil. Some sites contain 'made ground' or 'fill', which is material put there by man. Fill is unpredictable and unstable and may be contaminated. The topsoil (which contains all the living matter) is always removed down to the subsoil. When building on subsoil, it compresses, and the house moves. Foundations spread the load to minimize the movement and keep it consistent across the whole building. The object is to keep the movement so small that no significant damage is done to the house. In order to do this the foundations must be wide enough to spread the load adequately or, if the soil near the surface is too weak, they must be sunk deep enough to reach stronger soil lower down. Subsoils are of the following broad types:

- **Cohesive** This consists of fine particles that tend to stick together, such as clays and silts.
- **Non-cohesive** This consists of larger particles that tend to flow, such as sands and gravels.

Subsoils are often a mixture of the two, and are described using such terms as 'sandy clay'. Often there are larger grains like rocks or pebbles mixed in. Another soil type you may encounter is peat. This type of soil is extremely weak. Cohesive soil expands and shrinks as it wets and dries, because the fine grains draw closer together when water is lost from between them. The amount of water in the soil is affected by rainfall and by vegetation, and goes through an annual cycle. Usually, only the upper levels of soil (down to around 900 mm) are affected by water gain and loss due to the weather. However, trees affect the water content of much deeper soil, drying it out. Planting or removing a tree can cause the water content of the soil to change over a period of years. Wet soil will expand if frozen, but this affects only the top 450 mm of the soil on an average unexposed site.

CHOOSING TYPES OF FOUNDATIONS

See pages 122–126 for a comprehensive guide to the pros and cons of different types of foundations.

CONNECTING TO THE SERVICES

When it comes to connecting to the various services, in general the company that maintains the network will connect you to the system. They will all make the connection themselves, installing a branch towards your property. They employ contractors, with whom you may have to deal directly in order to coordinate the actual site work, and it will be cheaper if you do the groundwork yourself, digging the necessary trenches.

The secret of success is planning. As connecting the services involves potentially disruptive excavations both inside and just outside your site, you need to work them into your programme early on. It makes sense to lay ducts while you still have the heavy earth-moving equipment on site. Therefore, you need to get the applications in early. Utility companies are increasingly asking for more information, such as predicted loads, as their systems come under more strain. It might take some time to put this together, and you might need early advice from your

It makes sense to lay ducts for services while you still have the heavy earth-moving equipment on site.

tradespeople when filling in the forms. Allow good time for the applications to be processed so that any glitches can be ironed out. Some companies give lengths of ducting for free; other pieces, like meter enclosures, you have to buy. Finally, make allowances for the fact that the utilities and their contractors are often very busy.

- **Water** A pipe is run by the water company to an underground meter near the boundary of your site, usually just outside it. This is connected, usually by a plumber, to the house via a pipe. The water company will probably want to inspect the pipe and may want to check the entire plumbing system, including the choice of appliances.
- **Gas** The gas company will lay a pipe all the way to the house, and to a meter in a hatch on the house wall or next to it. The building inspector will vet the gas installations inside the house, by requiring them to be certified by a CORGI-registered plumber (Council for Registered Gas Installers).
- **Electricity** The situation is similar to gas, and again the building inspector will want to see a certificate issued by an NICEIC-registered electrician (National Inspection Council for Electrical Installation Contracting).
- **Telecommunications** Companies will connect you from overhead or via a duct.

ALTERNATIVES TO THE REGULAR UTILITIES

If you don't have a piped supply of water nearby, you may consider a borehole to extract water from the ground. The Environment Agency should be consulted as early as possible and will advise you on the practicalities of this, such as licensing and whether the water is drinkable. You can save on water with a recycling system and a purely gravity-fed system (like a water butt) can save a very worthwhile amount of water even if only for garden watering, or cleaning.

There is no law against extracting heat from the ground, and heat pumps are often an economic alternative to an expensive gas connection. Ground-source heat pumps are commonplace in Scandinavia and not uncommon in remote parts of the UK. They consist of a very long loop of buried pipe filled with water, which absorbs heat from the

WHO'S WHO?

Utility connections have become more complex with privatization and the opening up of markets to competition. Here's a quick summary of who does what:

- For gas and electricity there is a company that maintains the network (or 'infrastructure') and delivers the energy, and a choice of suppliers (who sell you the energy).
- For water supply the same company maintains the network and sells you the water.
- For telecommunications one company maintains most of the network, and you can 'buy' your calls from them or from their competitors; meanwhile, rival networks offering slightly different services have appeared.
- The drainage/sewerage system is owned by a company, which charges you to use the system.

ground as it circulates. A device, not unlike a refrigerator acting in reverse, 'upgrades' heat from this water to a small amount of warm water. This water is generally used for underfloor heating, but it is warm enough for other uses too. Other obvious alternative sources of heat are wood- and coal-burning appliances, which are now efficient enough to be allowed in smokeless zones. Multi-fuel stoves give flexibility, while wood-pellet stoves are seen as an environmentally friendly option, although there are not many suppliers of the fuel (made from waste wood) yet, and its price may go up or down in future.

Alternative renewable electricity supplies are becoming cheaper and more efficient, and a well-thought-out system of batteries, solar photovoltaic panels and/or wind turbine may ensure a reasonably reliable supply to what would have to be very efficient direct current (DC) appliances (see pages 188–192).

SEWAGE DISPOSAL

Where your plot is sited, and the type of land it is on, will determine what type of sewage disposal you will need to adopt. Even if a connection to the main sewage system is possible, you may prefer not to use it, although this is usually by far the cheapest option. The alternatives on offer are septic tanks, package treatment plants, cesspits or reed-beds in the grounds of the house. Each of these will need to be vetted by the Environment Agency. Your local building control officer will be able to advise you on what system would be most suitable.

For systems independent of the main sewerage network, surface water or rainwater should go to a natural watercourse, or a soakaway if the ground is suitable. The Environment Agency will take an interest in any non-sewer drainage systems, and should be consulted early. The agency has regulatory powers and can enforce the system you should use. It also holds a good deal of useful information and is a good source of advice as to what will work.

The drainage of 'foul' water (from baths, showers, WCs) is kept separate from surface or storm water (rainwater from roofs and paving), to avoid the foul-water system being overloaded. Some areas have combined drainage systems, in which the foul and surface water all go into one pipe. You will still be required to keep them separate on your plot until the last manhole, from which one pipe can take all the discharge to the public sewer.

Your local water/sewerage company, or utility, will have plans of its sewerage system. These plans are often incomplete and if you suspect there is a public sewer nearby that is not on the plan, inform the organization and it should inspect it. A public sewer on private land is owned by the utility, but you would need the landowner's consent to connect to it.

You have to apply for the utility's permission if you want to build over or near a public sewer. Building over may be prohibited if the utility considers that this will cause serious maintenance issues. It may insist that it is diverted at your expense, if it is practical to do so. You will be charged a deposit to cover any damage to the sewer.

Walls that work

A good architect or package house provider can be relied on to design walls that function well and pass Building Regulations. However, it's worth knowing how this component of your house works, as there are many choices available that can enhance or personalize it. House walling in the UK is dominated by two quite different technologies – timber-frame and masonry. Today we use synthetic or processed masonry – concrete or brick rather than the more traditional stone – and softwood rather than native

---

GETTING THE SERVICES RIGHT

- The most expensive aspects of utility connections are the digging up and re-laying of roads and pavements. If the services aren't in the pavement on your side of the road, the cost can run into thousands of pounds.
- You will require a water supply for the build. An electricity supply is also useful to avoid the noise, smell and expense of generators. A telephone connection will save a fortune on mobile-phone bills and a fax machine can be a bonus.
- Generally, a water supply for use on site can be applied for on the same form as the permanent supply, and the meter at the site boundary can be installed early and used for both.
- Consider using a larger bore water pipe from the meter to your house if it's a long way and you want a good flow rate because the friction of water passing through a long, narrow pipe can restrict the flow. However, too large a pipe can suffer

turbulence that again will slow the flow down. You should check with the water company and your plumber.
- There is a limit to how far the electricity network utility will run its cable into a site, perhaps 20 m. If your house is a long way from the boundary you may have to erect a meter enclosure partway and run your own cable from there to the house.
- With electricity you can arrange for a temporary supply, for which you must provide a weatherproof enclosure into which the infrastructure company will take their cable and install the service head. You can get the permanent supply installed once the permanent meter enclosure is ready, which may be some way into the build. The utility cannot physically move a service from a temporary to a permanent location. The service has to be cut off, the cable re-laid and the service head reinstalled.

hardwood in our frames. During the last few decades, increasing concern for the environment has caused something of a revolution, and both construction methods have been revised to incorporate increasingly higher levels of insulation (see page 119). This, and the desire for quicker building construction, has helped increase the interest in new types of walling, many of which blur the distinction between masonry and framing. These include polystyrene concrete-filled blocks, structural insulated panels (SIPs), prefabricated modular building, and other factory-made panel systems. Furthermore, there has been renewed interest in natural low-impact materials – native oak, earth, straw-and-lime – and a revival of the ancient skills applied in their construction.

EXTERNAL WALLS

Your external walls must be considered carefully, as they usually have to do all of the following things:
• carry structural loads
• resist wind loads
• resist rain and snow
• resist fire
• resist noise
• resist damage
• be durable
• provide security
• keep in heat
• be affordable
• be cheap, easy and safe to maintain
• have the appearance you want.

In addition you might want them to do the following:
• be environmentally friendly
• store heat in their thermal mass
• be quick/easy/satisfying to build on your particular site and by the labour available
• dry out quickly after construction
• conceal wiring and pipes easily.

How external walls work: structure

Walls have to carry loads. In a cavity wall, the structural strength is in the inner layer of blocks

The exterior of your house must fulfil a number of criteria, primarily protection against the elements, but also resistance to fire, noise and heat loss.

(although the outer skin contributes to stability). The blocks are made in various densities, the more dense they are the stronger they are. Lighter blocks are typically made of Aircrete – a mixture that includes cement. Usually they are laid in 10-mm thick mortar joints, although a new thin-joint system allows the mortar mix to be between 2 and 3 mm.

In a timber-frame wall the frame provides the strength. Timber-frames are usually made up of softwood sections (typically either 90 mm or 140 mm thick). The vertical pieces are called 'studs' and the framing is often referred to as 'studwork'. The more closely spaced, or bigger, the studs are, the stronger the wall. Boards called 'sheathing' fixed to the face of the frame give it stiffness or 'racking strength'.

Internal lining or coating

A lining is a board material, while a coating is a wet material that sets, such as plaster. They hide the internal structure of the external wall, and create a flat surface that can be decorated. A coating on a masonry wall helps to keep the wall airtight by filling the tiny cracks in the blockwork. Coating or lining a timber-frame is essential to protect it from fire and the spread of fire over its surface. Coatings and linings also help support attachments such as kitchen units, although heavy objects should be fixed through to the structure. There are many types of lining board, of which plasterboard – a sandwich of paper and plaster – is the most common. Basic plasterboard is referred to as 'wallboard'. Variations include the following: moisture-resisting, for use in steamy environments; foil-backed boards, which prevent vapour from passing through the board; tile-backer boards, which are strong and waterproof, and used for fixing tiles in wet areas (kitchens and bathrooms); and thermal laminates, which consist of a layer of board fixed to a layer of insulation, and used on masonry or timber-frame constructions, providing both a lining and extra insulation. Numerous other types differ in weight, thickness and flexibility for better sound resistance, holding of fixings, or to form curves.

Plaster is the most common type of coating and, again, there are various types for different backgrounds (the surfaces to which the plaster is applied) and with a choice of densities. For example, heavier plasters are stronger and help more with

soundproofing. Some are applied as a single coat but it is more usual to use two. Coatings can be applied to frames with 'lath' – wooden strips – although it is now more common to have expanded sheet metal or wire mesh.

Lining boards have a different appearance from plaster, with joints showing where the boards meet. A thin skim coat of wet plaster can be put on some lining boards to achieve the look of plaster. The alternative is to cover joints with a tape, which is quicker, drier and cheaper, but the joints will show if the work is not done to a high standard. Skim-coating is equally reliant on good workmanship to achieve a smooth and flat finish.

External wall finish: cladding

This is the wall's main protection from the weather. It also gives a house its appearance.
The three common types of cladding are:

- An 'external leaf' that stands in front of the main structure and supports its own weight. This is the outer part of a cavity wall, or the brick skin over a timber-frame. It is usually of brickwork, or rendered concrete blockwork. In principle it may not be 100 per cent waterproof, but the cavity prevents it damaging the structure.
- A 'rainscreen', which keeps water out on exactly the same principle as an external leaf, but is thinner and lighter and hangs off the structure. Common rainscreens are render-on-board, tile-hanging, timber boarding or synthetic boarding. Each is fixed to timber battens attached to the main structure. Many of these, even including some types of timber, are as maintenance free as brickwork.
- An 'external wall insulation system', which is synthetic render over rigid insulation fixed to the structure or to an external leaf with no gap between the two. These are quite new to the UK market. As there is no cavity behind, they are completely reliant on being 100 per cent waterproof to prevent water getting to the structure. This is achieved by the render being flexible and reinforced (quite different from traditional sand/cement/lime render). The render can be self-coloured, so as to not need re-painting. See pages 147–149 for detailed information on insulation.

Other types of external wall

Some new walling types incorporate structure and insulation in one. For example SIPs (see page 106) take the load and have rigid insulation. Kingspan's TEK Building System is a sophisticated variation of this system for a whole-house design (see page 106). Alternatively, BecoWallform comprises hollow blocks, which act as a mould for a concrete wall that incorporates both structure and insulation. Environmentally friendly options include straw-bale and earth or mud walling. These can make wonderful walls but need care both in construction and maintenance (see pages 192–195).

Better external walls

Understanding the function of each layer helps to understand how an external wall can be improved.
• Sound resistance can usually be enhanced through the use of heavier materials, such as denser masonry or linings, and by making sure all gaps are sealed. It is worth noting that doors and windows are always less sound resistant than walls, so there is a limit to what you can achieve here.
• Walls can be more environmentally friendly, through the careful selection of materials if not of the building system as a whole.
• Walls can be easier to build using small, light components or easy construction systems. The former would be especially important on a restricted site that a crane couldn't reach.
• More, or better insulation, will reduce bills and environmental harm.
• Using dry systems (avoiding mortar and plaster) might be important on a fast build or where you are fitting moisture-sensitive finishes to a complex building and you don't want to have to wait for the building to dry out (which can take weeks).
• Fast or prefabricated systems are essential if a quick build is important.
• If your site is south-facing you might choose to build a thermally massive house to soak up the sun.
• Thinner walls will allow more living space on a restricted site – it is surprising how much space is gained by using a slightly thinner wall.

Choosing external walls

See pages 127–130 for a comprehensive guide to the pros and cons of different types of external walls.

INTERNAL WALLS

Your internal walls usually have to do all of the following things:
• carry structural loads
• contribute to resisting wind loads
• resist fire
• resist noise
• resist damage
• be durable
• be affordable
• be easy to maintain
• have the appearance you want.

In addition you might want them to do the following:
• be environmentally friendly
• store heat in their thermal mass (see page 194)
• be quick/easy/satisfying to build, on your particular site, and by the labour available
• dry out quickly after construction
• conceal wiring and pipes easily.

How internal walls work: structure

Masonry blocks and timber-frames take the loads. The linings or coatings perform the same functions as they do on an external wall (see below). They resist noise, too, by adding mass to the wall. A coating on masonry helps with soundproofing by covering over cracks in the joints, while linings keep timber partitions rigid. Internal walls can always be made more soundproof with denser blocks, coatings and linings, and by good workmanship, paying attention to eliminating gaps. Natural coatings such

Timber tends to shrink over time and adequate insulation needs to be installed at the time of building the internal walls to prevent gaps forming.

Here, the beams of a beam-and-block floor have been laid ready for the blocks to be rested between them.

as lime and clay plasters are more environmentally friendly and can make softer, more tactile finishes possible. There are some natural paints on the market, which are kinder to the environment and release fewer fumes into the atmosphere than more conventional products.

Other types of internal walls

Some products are commonplace in mainland Europe, but have yet to gain a foothold in the UK market. To give you an idea of some of the types available and the advantages and disadvantages of each look at the table on pages 131–132. The Steko wooden block system, for example, makes use of softwood offcuts to make a strong, dense wall that can be assembled without adhesives or fixings. Timbatec stud framing is said to be more soundproof, while Karphosit uses clay blocks laid in a clay 'mortar' and is among the most environmentally friendly options available.

Lining or coating

As with products for an external wall, lining is a board material while a coating is a wet material, such as plaster, which sets to form a flat surface that can be decorated. Linings and coatings should resist the spread of fire over their surface and protect the structure of a timber-frame. They also need to support attachments such as light fittings.

Choosing internal walls

See pages 131–132 for a comprehensive guide to the pros and cons of different types of internal walls.

Floors that work

Buildings can have two types of floors: ground floors and upper floors.

GROUND FLOORS

Your ground floors usually have to do all of the following things:
• provide a flat, level surface
• keep dampness out
• keep heat in
• be affordable
• have the appearance you want.

In addition you might want your ground floor to:
• support loads carried down by walls and columns
• be environmentally friendly
• store heat in its thermal mass
• be quick/easy/satisfying to build
• dry out quickly after construction
• conceal wiring and pipes easily
• accommodate underfloor heating.

How ground floors work: structure

Structurally, there are broadly two main types of ground floor: ground-bearing or suspended.

Ground-bearing floors rest directly on the ground, as the name suggests, while suspended floors bridge the foundations or walls. The latter is used if the ground is too unstable, weak or prone to movement to support a ground-bearing floor. A sloping site might need a lot of cutting and filling to create platforms for a ground-bearing floor, and it is not usually permitted to put a ground-bearing floor on fill deeper than 600 mm. To an extent your choice of ground floor is dictated by the site (see foundation types, page 108).

Ground-bearing floors are almost always concrete, while there are many types of suspended floor: concrete, steel and timber, for example. Enthusiasts for natural materials have been experimenting with old technologies like earth and clay, hardened and stabilized with blood and hair. A more likely option in the future is lightweight steel floor framing. The joists are C-shaped sections made of very thin steel. They are lighter, stiffer and more accurate than solid timber, but need to be wrapped with insulation to avoid cold spots and condensation. Currently used for large building developments, this is a relatively expensive option for the self-builder.

Many precast reinforced-concrete floors are 'cambered' (arched slightly). The camber can vary slightly between units, or can be up to 25 mm over a 5-metre span, so you may need a levelling screed (a non-structural mixture of sand and cement laid on top before putting down a floor finish). It is generally possible to put any kind of finish on any type of floor, although a timber sub-floor might need to be built very strongly to take something as heavy as flagstones.

Damp-proofing the ground floor

Vapour control, damp-proofing and ventilation need to be designed into any floor carefully. Moisture comes from the ground, from inside the house, and from the materials themselves (all materials dry out after construction).

There are various options as to where the membrane is placed in a construction, and specialist advice should be sought: omitted or used wrongly, inadequate damp-proofing can result in condensation, rot, cold spots and mould growth. (See pages 147–149 for detailed information on insulation.)

Environmentally friendly ground floors

Ground floors can be more environmentally friendly if fill and hardcore are recycled concrete rather than new aggregates; if timbers are from sustainable forests and not imported over great distances; and if insulation is carefully selected. At the very least ground floors can always be better insulated by using more powerful insulation or by increasing its thickness.

Choosing ground floors

See pages 133–137 for a comprehensive guide to the pros and cons of different types of ground floors.

UPPER FLOORS

Upper floors have to do the following:
• provide a flat, level surface
• span between walls without deflecting (bending)
• resist passage of sound
• be affordable
• have the appearance you want.

In addition you might want your upper floors to do the following:
• support loads carried down by walls and columns
• be environmentally friendly
• store heat in their thermal mass
• be quick/easy/satisfying to build
• dry out quickly after construction
• conceal wiring and pipes easily.

How upper floors work: structure

Upper floors have to be strong enough to bridge between walls without breaking or bending excessively. They also have to comply with Building Regulations on sound insulation for internal house floors. Any floor can be made more soundproof in the following ways:
• resilient coverings such as carpet, or more elaborate foam-backed boards (these will help particularly with impact sound)
• more mass, such as thicker or denser concrete, or thicker or heavier linings and coatings
• suspended ceiling linings, the more 'detached' the better. The ultimate is an entirely separate, self-supporting ceiling.

If sound gets through it is due to poor detailing or workmanship. It is important to make absolutely

There are many types of floors available with varying degrees of sound resistance.

Finishes for upper floors

Materials are laid over the structure, giving a flat surface to walk on. Timber floors usually have timber-based board materials. Concrete floors may need a screed to flatten out any camber (see ground floors) or grout to fill up gaps. On top of these go the final finishes of your choice, such as carpet or tiles.

Wet rooms

Wet rooms need to be finished carefully to avoid the risk of water leaking into the building structure. Regulations now demand that water-resistant board materials are used for wet rooms. The edges of the floor – where shrinkage gaps can open up – are the weakest points. Sophisticated tanking fluid system with reinforced edges can usually achieve a water-tight room as long as the walls and floor are reasonably stiff and rigid.

Choosing upper floors

See pages 138–140 for a comprehensive guide to the pros and cons of different types of upper floors.

sure that the floors are airtight and, because materials shrink, caulk them with something that expands. Where safety is concerned, there are obvious risks involved with handling structural materials at upper-floor level and, in particular, no one should be allowed below when a beam-and-block floor is being assembled.

One of the benefits of building with a timber-frame is that there is no need for internal walls. This means there is plenty of scope for large internal spaces and vaulted ceilings.

Flat roofs have a reputation for being unreliable and yet there is no reason why they cannot perform as well, and for as long, as pitched roofs.

Roofs

Big, pitched roofs are a prominent feature of houses in our wet and potentially snowy climate. However, flat roofs have been a part of the scene for a long time too – at least as far back as Elizabethan times, when nobles would take the air on lead-covered roof terraces. Pitched roofs have the advantage of space inside them. If not used for storage, this can become a living space. It is a way of sneaking an extra (albeit smaller) storey of living space into a house without increasing the impression of size too much. This can also be a relatively cheap extra space.

The fashion for flat roofs in the early 20th century showed up the limitations of the technology of the time. Relatively recently, advances in materials have made it possible to build inexpensive, yet quite durable flat roofs. They still rely considerably on good design and construction to avoid premature failure, and the durability of the most promising

Copper has been used to create a contemporary-looking roof structure for this house. The green of the roof offers a striking contrast to the red brick of the walls.

materials is uncertain simply because they have not been in use for very long.

One consideration when choosing a roof is to opt for one that works with the style of your house. Each age had its favoured style, pitch, features and materials and the wrong one could look very odd. For instance, a roof terrace on a medieval oak frame or an Arts and Crafts style house will simply look out of place.

DURABILITY AND MAINTENANCE

The common conception that pitched roofs require no maintenance while flat ones need periodic replacement is overly simplistic. The most durable roofs are in fact the metals, and they can be laid to a pitch as low as 5 degrees. Nobody knows how long flat single-ply membrane roofs will last, because the oldest ones are under 50 years old (although they are showing no signs of failure). Asphalt roofs can last for over 50 years if protected from the sun, and serviceable, 100-year-old asphalt roofs have been known. It is recommended that built-up roofs are replaced periodically to guarantee watertightness. If flat roofs fail it is because they are laid badly, designed badly or walked on when they shouldn't be. Pitched roofs tend to fail owing to lack of maintenance or the timber supporting the battens rotting (through leaks during storm damage, for instance).

All roofs should be looked after, and a little attention will prolong life. It's a good policy to inspect any roof every few years and repair or replace as necessary. Paving flat roofs for use as terraces can prolong their life by protecting them from the sun and mechanical damage; on the other hand, badly laid paving can have the opposite effect, overloading the roof and causing it to sag, so that water pools. Furthermore, without the proper mountings or protective layers, paving can dig into the roof finish and puncture it. Roofs tend to fail at the junctions around the edges, and paving may not protect these.

It's useful to understand that most pitched-roof finishes, the ones made up of small overlapping units like tiles, rely on a second layer, the 'underlay', because the tile layer is not 100 per cent waterproof. Flat and metal roof finishes, however, are designed to keep all the water out.

Mineral fibre insulation is supplied as a rolled-up soft slab. It is easy to cut and fit and is installed compressed so that it expands to fill gaps.

Choosing roofs

See pages 141–146 for a comprehensive guide to the pros and cons of different forms of roofs.

The right insulation

By insulating your home correctly you'll enjoy greater comfort and reduce your heating bills.

FIBROUS INSULATION

Soft fibrous insulation, such as mineral fibre or cellulose fibre, is often used in timber-framed walls, floors and roofs because it can be cut slightly large and pressed into the gaps. As the timber shrinks or moves slightly the insulation takes the new shape while remaining a tight fit. Also, small pieces can easily be cut and packed into awkward-shaped areas.

Cellulose fibre, made from recycled newspaper, can be sprayed dry into walls, floors and roofs or on to ceilings in attics in what is known as a 'breathing' or enhanced vapour transfer (EVT) timber-framed construction, which allows vapour to diffuse naturally through it. Cellulose fibre can be sprayed slightly damp, in which case it sets like a cake and supports itself to some extent until it is held in permanently. Lining board is then attached. This uses cellulose fibre's ability to absorb a certain amount of moisture and then dry out without permanent damage. It is believed to be good for timber-frames because it helps to dry out any moisture that might get into the wall, by 'wicking' it away from the timber and drying it out to the atmosphere.

Other less commonly used products are more expensive but have particular qualities or advantages. Many of them, such as sheep's wool, are environmentally friendly in their use of natural products, in contrast to the plastics-based foam boards, which are manufactured.

INSULATING BOARDS
Boards, most of which are plastics based but include cork and foamed glass, are suited for use, butt-jointed into a continuous surface, such as in masonry cavity walls, in concrete floors, or fixed to the face of framed structures. Most of them are rigid enough to support loads, which is why they are used in floors. To varying degrees they are waterproof and so can be used in exposed or damp situations such as in some types of flat roof or in foundations.

Boards are being used increasingly within framed structures: they are more powerful than fibrous insulation for the same amount of space they take up. Boards should be cut accurately to size and well-dried timber should be used, otherwise gaps caused by poor fit or timber shrinkage may result in cold spots and possibly condensation in the structure.

Insulation materials vary in their 'power' or as to how good they are at keeping heat inside the house and the cold out. For the same amount of space that they take up, phenolic board is about twice as powerful as fibre glass and all the other types fall between the two. Generally, for the mainstream products, the more powerful the insulation, the higher the price.

CHOOSING INSULATION
See pages 147–149 for a comprehensive guide to the pros and cons of different types of insulation.

Construction methods checklist

LOCATION AND PROPERTIES OF THE SITE

☐ Choose the construction method that best suits your plot, your house design and your pocket.

☐ Make sure your construction method and plans are both legal and viable before any work starts.

☐ Gain as much local knowledge as you can about the land in which your plot lies.

☐ Carry out a ground investigation.

USING THE RIGHT MATERIALS

☐ Find out what you need from your external walls above and beyond the essential weather-resisting and insulating properties.

☐ Check that any specialist materials and skills you will need for your external and internal walls are available in your area.

☐ Choose a roof to suit the look of your house design and be functional and consider all the options.

☐ Ensure that you have adequate insulation. The better insulated your home, the more money you will save on your heating bill.

☐ Equipment and materials can be dangerous, so make sure you have the right safety equipment and seek professional advice on safe practice.

FUTURE PROOFING

☐ Build in the option of a loft conversion now and it will add to the value of your house because the conversion will be cheaper and easier when it does happen.

FOUNDATION TYPES
Shallow Foundations

Type	What is it?	Where is it appropriate?	What types of ground floor are compatible?	What are the advantages?	What are the disadvantages?
Strip	A trench is dug (generally 600 mm* wide) and partly filled with concrete (to at least 150 mm deep). Walls are built on it up to ground level. The width is proportional to the load.	• All soils with a reasonable bearing capacity. • Very small buildings on inaccessible sites (the relatively small amount of concrete can be hand-mixed).	• Ground-bearing slab. • Suspended timber. • Beam and block. • Precast concrete.	• Lower materials cost. • Less concrete than most foundation types, so more environmentally friendly.	• Higher labour cost. • Creates awkward and potentially dangerous work in trenches.
Wide strip	A wide trench partly filled with concrete. The concrete has some reinforcement to prevent the parts beyond the wall from breaking away. Walls are built on it up to ground level.	• All soils with a reasonable bearing capacity; can be used for greater loads and/or weaker soils than narrower strip.	• Ground-bearing slab. • Suspended timber. • Beam and block. • Precast concrete.	• Lower materials cost. • Less concrete than most foundation types, so more environmentally friendly.	• Higher labour cost. • Creates awkward and potentially dangerous work in trenches.
Trench fill	A trench (typically 600 mm* wide x 900 mm deep) is made and filled with unreinforced concrete up to just below ground level.	• Cohesive soils where the upper layers are weak soil, fill or made ground.	• Ground-bearing slab. • Suspended timber. • Beam and block. • Precast concrete.	• Fast, inexpensive.	• Uses a lot of concrete so not environmentally friendly. • Trenches may require support.
Deep trench fill	As above but up to 3 m deep.	• Cohesive soils where the upper layers are weak soil, fill or made ground. • Close to trees.	• Ground-bearing slab. • Suspended timber. • Beam and block. • Precast concrete.	• Fast.	• Uses a lot of concrete so not environmentally friendly. • Trenches may require support.

Shallow Foundations (continued)

Type	What is it?	Where is it appropriate?	What types of ground floor are compatible?	What are the advantages?	What are the disadvantages?
Pad	Pits partially filled with concrete and 'piers' (columns) built up from them up to ground level. Ground beams span between the pads or piers to support the walls and floor.	• Cohesive soils where the upper layers are weak soil or made ground. • Close to trees. • For buildings with 'point loads'.	• Ground-bearing slab. • Suspended timber. • Beam and block. • Precast concrete.	• Less excavation than for strip and trench fill, also less concrete, so more environmentally friendly (pad more so than deep pad.)	• Beams need engineering design. • Creates awkward and potentially dangerous work in pits.
Deep pad	Pits filled with concrete. Ground beams span between the pads or piers to support the walls and floor.	• Cohesive soils where the upper layers are weak soil or made ground. • Close to trees. • For buildings with 'point loads'.	• Ground-bearing slab. • Suspended timber. • Beam and block. • Precast concrete.	• Less excavation than for strip and trench fill, also less concrete, so more environmentally friendly (pad more so than deep pad.)	• Beams need engineering design. • Creates awkward and potentially dangerous work in pits.
Flat raft	The ground is levelled and consolidated with compacted hardcore. A flat concrete slab (which may or may not be reinforced) is cast on the consolidated ground.	• Soil with poor bearing capacity. • Peat. • Made ground.	• Integral with raft.	• Combines functions of floor and foundation. • Provides working platform early in build.	• Uses a lot of concrete. • Requires careful pre-planning of service penetrations through the raft. • Requires engineering design.

NB *Trenches are typically dug to 600 mm wide to take the weight of an outside wall of the house. Trenches that are dug to 450 mm wide are considered enough to take the load of an interior wall – providing it is non load-bearing.*

Shallow Foundations (continued)					
Type	**What is it?**	**Where is it appropriate?**	**What types of ground floor are compatible?**	**What are the advantages?**	**What are the disadvantages?**
Raft with down-stand	As previous but the reinforced concrete raft is thickened downwards at edges and reinforced to form an 'edge beam', and thickened under internal load-bearing walls.	• Soil with poor bearing capacity. • Peat. • Made ground. • The downstand (downward folding edge) allows better building on weaker soils rather than flat raft.	• Integral with raft	• Combines functions of floor and foundation. • Provides working platform early in build.	• Uses a lot of concrete. • Requires careful pre-planning of service penetrations through the raft. • Requires engineering design.

NB *Trenches are typically dug to 600 mm wide to take the weight of an outside wall of the house. Trenches that are dug to 450 mm wide are considered enough to take the load of an interior wall – providing it is non load-bearing.*

Deep Foundations

Type	What is it?	Where is it appropriate?	What types of ground floor are compatible?	What are the advantages?	What are the disadvantages?
Bored pile	Deep shafts are drilled and filled with reinforced concrete.	• Where upper layers of ground are too weak. • Sites close to trees, or with a high water table and poorly drained. • Peat. • Made ground.	• Ground beams can span between the piles to support a suspended timber, beam-and-block, precast concrete or in-situ concrete floor. • An in-situ reinforced concrete suspended flat slab can be cast across the tops of the piles.	• Less excavation. • Shallow structure – less damage to trees, especially with a flat slab.	• Piling rig may require ground stabilization to provide a working platform. • Expensive specialist work.
Driven pile	Prefabricated reinforced concrete piles are driven into the ground.	• Where upper layers of ground are too weak. • Sites close to trees, or with a high water table and poorly drained. • Peat. • Made ground.	• Ground beams can span between the piles to support a suspended timber, beam-and-block, precast concrete or in-situ concrete floor. • An in-situ reinforced concrete suspended flat slab can be cast across the tops of the piles.	• Less excavation. • Shallow structure – less damage to trees, especially with a flat slab.	• As above. • Vibration risk to nearby structures.

Other foundation types					
Type	**What is it?**	**Where is it appropriate?**	**What types of ground floor are compatible?**	**What are the advantages?**	**What are the disadvantages?**
Vibro compaction, vibro replacement or grout injection	A probe is lowered into the ground and compacts the loose ground by vibrating (compaction), injecting aggregate into the voids left (replacement), or injecting grout (grout injection).	• Weak soils. • Made ground.	• The 'columns' of compacted ground can support pads and ground beams, which in turn can support a suspended floor.	• Less excavation.	• Vibration risk to nearby structures. • Expensive specialist work.

EXTERNAL WALL TYPES

Type	What is it?	What are the advantages?	What are the disadvantages?
Masonry cavity wall	Two 'skins' or 'leaves' of masonry with a cavity between. The inner skin is usually concrete block. The outer skin may be brickwork, rendered block or stone. The cavity can be fully filled with mineral fibre or partially filled with rigid board insulation. A coating or lining is applied to the inside surface of the inner skin. Any kind of lining can be used, including thermal laminate, or a coating.	• Durable (brickwork requires little maintenance on the outside; however, render requires painting). • The blockwork inner leaf is easy to fix heavy items to.	• Slow, but the process could be speeded up by using the thin-joint system. • Wet (using the thin-joints system and lining board rather than coating can improve this).
Timber stud frame with rainscreen	Softwood framing with rainscreen cladding attached to the outside and lining to the inside. The framing can be assembled on site (stick-built) or prefabricated as panels. Can be part of a whole house prefabricated kit in which the wall panels can be supplied already insulated. Any kind of lightweight cladding can be used (subject to site exposure).	• Can be prefabricated, giving more precise and accurate walls and a fast build. • Light. • Quick, especially if prefabricated or modular. • Dry. • Can achieve a highly insulated, relatively thin wall.	• Hanging very heavy items needs preplanning to provide timbers (or a plywood layer) to coincide with fixings. (Some linings can take heavy fixings.) • Good workmanship is important for achieving durability.

EXTERNAL WALL TYPES (continued)

Type	What is it?	What are the advantages?	What are the disadvantages?
Timber stud frame with rainscreen (continued)	Board or mineral fibre insulation between the studs can be attached to the inside or outside of the frame or any combination of these. Usually lining boards inside.	• The frame can be an environmentally friendly option, especially if timber is from a sustainable source.	• Not all builders are experienced at timber-frame building.
Timber and brick	As above, but with a brickwork outer leaf instead of a rainscreen. It is also quite common to use an outer leaf of rendered blockwork. Board or mineral fibre insulation between the studs can be attached to the inside or outside of the frame or any combination of these. Usually lining boards are used inside.	• Combines the appearance of brick with the advantages of a timber-frame. • A weather-tight structure can be built quickly (the brickwork done later).	• As above • The foundation has to be more substantial to support the brick skin. • Needs careful detailing as timber and brick shrink and move differently.
Light gauge steel frame	Very much like a timber-frame but using thin, light, folded steel sections.	• Many of the advantages of timber-framing without the shrinkage. • Can be used with brick outer leaf or rainscreen.	• The same as timber-frame, plus: careful detailing needed to avoid cold bridges and condensation. • Unconventional building. Suppliers are not set up for one-off houses and this increases the cost.
Kingspan's TEK Building System	A variation on the structural insulation panels (SIPs) for whole-house construction. Factory-made panels comprising high-performance insulation	• Ultra-efficient (thin, well-insulated wall). • Very fast construction.	• Limited range of standard designs. Can be ordered to your own design.

EXTERNAL WALL TYPES (continued)

Type	What is it?	What are the advantages?	What are the disadvantages?
Kingspan's TEK Building System (continued)	High-performance insulation core with wood-based board. They are assembled into houses without requiring additional framing. Any rainscreen or outer leaf can be used on the outside. Any kind of lining board can be used, spaced off from the panel to provide space for services (you cannot run services in the panel).	• Accurate, factory-made product. • Very light.	• A crane and good access to site are needed.
LenoTec	Prefabricated laminated softwood panels, which are self supporting. Any kind of rainscreen or outer leaf. Thermal lining is used to achieve acceptable level of insulation.	• Can be ordered to your own design. • Very fast construction. • Very stable. • No shrinkage. • 'Breathing' construction.	• Heavy, so needs crane and heavy lifting equipment and good access to site. • Uncommon.
Oak frame	Oak columns and beams pegged together with infill. Any kind of rainscreen or outer leaf. When the frame is exposed on the inside and outside, the infill is a special combination of cladding, insulation and lining. Often the frame is hidden, either on the outside by rainscreen or on the inside by lining boards.	• Environmentally friendly. • Beautiful effects possible. • Promotes broadleaf forestry and traditional craft skills.	• Oak shrinks, making airtight capacity difficult and necessitating complicated movement joints and seals that will need periodic checking.

EXTERNAL WALL TYPES (continued)

Type	What is it?	What are the advantages?	What are the disadvantages?
Polystyrene block	Hollow interlocking polystyrene blocks, the hollows filled with reinforcement and concrete. Comes as a system with lintels for openings and brackets for supporting floors. External finish is rendered directly on to the blocks or rainscreen. Internal finish is rendered directly on to the blocks.	• Suitable for DIY use.	• Uncommon. • Block module places some limitations on dimensions. • Uses a lot of concrete.
Straw bale	Straw bales are stacked up. Can be load bearing or inserted into a post-and-beam frame. 'Breathable' render inside and outside (usually lime or clay based). Rainscreen cladding is an option.	• Environmentally friendly. • Well insulated. • Beautiful 'soft' shapes and finishes possible.	• Uncommon. • Prone to durability problems unless detailed carefully and render periodically inspected and well maintained. • Walls are very thick. • Has to be designed to bale module.
Log	Basically, logs stacked up – but there are many variations, some quite high tech: logs are machined almost square, laminated, given tongue-and-groove joints, doubled up to form cavity walls, dry lined, and jointed with non-setting mastic. Sometimes no further finishes or linings are applied inside or outside, but thermal lining may be necessary to provide adequate insulation.	• Beautiful effects possible. • Environmentally friendly.	• Difficult to achieve good insulation and airtightness, particularly if you want to be able to see the logs on the inside. • It can be difficult to get planning permission for log houses.

NB *Items marked 'uncommon' or 'unconventional' may have the following disadvantages: limited number of suppliers (not mainstream merchants), so accurate assessment of quantities and pre-ordering required; relatively expensive due to being a small market; builders overcharging to cover the risk of using unfamiliar materials; difficulties with Building Control approval.*

INTERNAL WALL TYPES

Type	What is it?	What are the advantages?	What are the disadvantages?
Softwood 'stud' frame	Softwood rectangular sections nailed together to form framing. Load bearing (limited strength – usually supporting timber floors only). Lining boards are usually applied to both sides, although coating on lath can be used. Mineral fibre or foam board insulation can be added between the framing.	• Fast and dry. • Easy to run wiring and pipes through it.	• Hanging very heavy items needs preplanning (extra timbers in the structure; some strong lining boards take quite heavy loads).
Concrete blocks	Concrete blocks laid in mortar. Load bearing (strength varies with weight or density of blocks). Coatings are usually applied to both sides but lining boards can be used.	• Easy to fix heavy items to.	• Wet (the mortar needs drying out, and so do coatings if they are used). • Cannot be built on a timber floor.
Light gauge steel framing	Very much like a timber frame but using thin, light, folded steel sections. The same linings or coatings and insulation are used. Load bearing.	• Fast and dry. • Easy to run wiring and pipes through it. • More accurate than softwood.	• Uncommon.

INTERNAL WALL TYPES (continued)

Type	What is it?	What are the advantages?	What are the disadvantages?
Karphosit clay blocks	Compressed clay/straw fibre blocks used with a clay mortar. Clay, lime or gypsum plaster can be applied. Non load-bearing.	• Pleasant to use. • Said to improve air quality. • Can be used with clay plasters for a soft finish. • Environmentally friendly.	• Uncommon.
Glazing	Glass in various types of framing, usually softwood. Non load-bearing.	• Beautiful effects possible.	• Poor sound resistance.
Glass blocks	Hollow glass blocks laid in special mortar and with special reinforcement and brackets. Non load-bearing.	• Beautiful effects possible. • Heavier ones have good sound resistance.	• Installing them properly can be quite complex. • Heavy. • Requires rigid support.
Steko blocks	Interlocking wooden blocks stacked up. No glue or fixings required. Load-bearing.	• Easy to use. • Environmentally friendly.	• Uncommon.
Timbatec stud framing	Timber studwork, but with studs made up of three timbers pegged together. Load bearing. Tends to be used in conjunction with clay lining boards and natural-fibre insulation.	• Greater acoustic isolation than ordinary studs.	• Uncommon.

NB *Items marked 'uncommon' may have the following disadvantages: limited number of suppliers (not mainstream merchants), so accurate assessment of quantities and pre-ordering required; relatively expensive due to being a small market; builders overcharging to cover the risk of using unfamiliar materials; difficulties with Building Control approval.*

GROUND FLOOR TYPES
Ground-bearing floors

Type	What is it?	What are the advantages?	What are the disadvantages?
Ground-bearing slab with insulation underneath	A reinforced-concrete slab, set inside the walls, on top of rigid, waterproof insulation board and laid on 'blinding' (a layer of sand or concrete to provide a level surface), The blinding is on top of hardcore compacted directly into the ground.	• Creates a potentially thermally massive floor because the concrete slab is on top of the insulation. • Can provide a durable working platform; alternatively, it can be installed after the walls have been built. • Service penetrations through the floor are easy and made by casting a sleeve for wires and pipes into the concrete while the concrete is wet. This needs preplanning.	• It is difficult to cast a concrete slab on top of insulation without damaging the insulation, so your floor is likely to be less well insulated than expected. • Need to get the concrete wagon close to the building (pumping is expensive). • Slab takes a long time to dry out. • A professional engineer, architect or technologist should be asked to advise on these kinds of floor.
Ground-bearing slab with insulation on top	A reinforced concrete slab cast on blinding (see above) on hardcore compacted into the ground. Rigid board insulation is placed on top of the slab, and on this, either a timber-board floor or a screed. A timber-board floor will be thermally lightweight, a screed will be heavyweight.	• Provides a durable working platform early in the build. • Service penetrations through the floor are easy (as above) but need preplanning.	• As above.

GROUND FLOOR TYPES
Ground-bearing floors (continued)

Type	What is it?	What are the advantages?	What are the disadvantages?
Raft	Like a ground-bearing slab with insulation on top, but acts as a foundation as well; the walls are built off the slab rather than the slab between the walls.	• This may be an option on bad ground. • Provides a durable working platform early in the build. • Combines functions of floor and foundation.	• Not very environmentally friendly because it uses a lot of concrete. • Requires careful preplanning of service penetrations through the raft, because the floor is a structural element and difficult to alter.

SUSPENDED FLOORS

Type	What is it?	What are the advantages?	What are the disadvantages?
Timber joist	Solid timber (usually softwood) joists bridge between the walls or the foundations. Decking (usually a timber-based board material) is fixed down to them. Insulation is placed between the joists. This can be a rigid board propped between them or a mineral fibre type supported on netting or board between the joists.	• The floor will be relatively thermally lightweight, although fairly thick and heavy tiles can be laid on a structurally upgraded floor. • For a given floor thickness, can provide by far the best levels of insulation. • Dry and fast to construct. • Made of light, easy-to-handle and easily available components. • Adaptable to changes, last-minute decisions and corrections. • Service penetrations easy. • Environmentally friendly, especially if timber from renewable source and not imported too far.	• The soil below has to be covered to prevent plant growth. • A deeper, drained underfloor void required. On a badly drained site this can make the ground floor rather high and make long ramps necessary for disabled access. • Prone to squeaking unless built from well-selected timbers and well constructed. • May be damaged if built early; temporary decking can be used to provide a working platform, or the floor can be built after the walls. • Spans may be limited to around 4.8 m unless the floor is split in two by a steel beam.

GROUND FLOOR TYPES
Suspended floors (continued)

Type	What is it?	What are the advantages?	What are the disadvantages?
Engineered timber	The same as a timber-joist floor but the joists are beams that are I-shaped in profile, made up of timber battens top and bottom and either a thin board or light steel struts between them to form the 'web' of the beam.	• The same as a timber-joist floor except that components have to be pre-ordered. • Straighter, lighter and more dimensionally accurate than solid timber. • Quite large pipes and ducts can be cut through the web of the beam or passed between the metal struts.	• The same as a timber-joist floor, except much less likely to squeak and capable of long spans. • More expensive. • They come as a system with their own detailing and fixings – tradesmen may not be familiar. • The beams and fixings need to be pre-ordered.
Beam and concrete block	Concrete beams shaped like an inverted 'T' with concrete blocks (sometimes referred to as 'pots') resting between them. The blocks can be ordinary wall building blocks, usually Aircrete (lightweight). A levelling concrete screed or topping may be needed on long spans to cover the camber. On top of this will be rigid board insulation and on this a floating timber floor (floorboards on battens) or screed.	• Provides a durable working platform. • Service penetrations through the floor are easy to achieve by leaving a block out. • Fast and dry (less so if a screed is applied). • Supplier will work out a layout and a schedule of suitable beams and blocks. They will supply only, or supply-and-fix.	• Beams need to be delivered close to their final position by the delivery truck or a small crane, as they are very heavy to lift. • This kind of floor seems deceptively simple, but corners and edges can be tricky, so a supply-and-fit contract with the manufacturer is recommended.

GROUND FLOOR TYPES
Suspended floors (continued)

Type	What is it?	What are the advantages?	What are the disadvantages?
Beam and polystyrene infill blocks	The same as beam and concrete block but specially shaped polystyrene blocks fit between the beams.	• A concrete screed is required to give the floor its structural strength. • As above but better insulated than the concrete block type.	• Less robust than concrete blocks during construction.
Hollowcore	These are concrete slabs, generally 1200 mm wide and 150 mm thick, with perforations through their length. They span between foundations in the same way as beam and block, and take the same insulation and thicknesses. They are cambered and so may require a levelling screed.	• Fast to install. • Provides a durable working platform. • Provides maximum structural efficiency. • Can be provided with long spans in excess of 16 m. • Dry (less so if a screed is applied).	• A crane close to the building is required. • Small holes can be drilled in them but large holes have to be cast into the plank in the factory, so you would need to plan your services carefully in advance. • Lead-in time for ordering. • Very difficult to adapt to changes, last-minute decisions and corrections. • Difficult to adapt to non-rectangular shapes.

GROUND FLOOR TYPES
Suspended floors (continued)

Type	What is it?	What are the advantages?	What are the disadvantages?
In-situ suspended floor slab with ground beams	Usually used with piles (supports) on bad ground (see page 125), this is a reinforced-concrete slab, spanning across reinforced-concrete ground beams between the piles. The slab is cast on temporary cardboard-based cellular support boards which dissolves away, leaving the slab spanning between the beams. The same finishes and insulation are used as for beam-and-block floors.	• Provides a durable working platform. • No limitation on shape, so good for odd-shaped buildings.	• Not environmentally friendly (uses a lot of concrete and reinforcement). • Need to get the concrete wagon close to the building (pumping is expensive). • You need to plan your services carefully in advance. • Very difficult to adapt to changes, last-minute decisions and corrections. • This kind of floor has to be designed by a professional engineer and installed by a main contractor or specialist subcontractor.
In-situ suspended flat slab	As above, but the beams are incorporated structurally into the concrete slab, resulting in a shallower structure overall.	• As above, plus the lack of ground beams mean you can avoid cutting through obstructions such as tree roots.	• As above.

UPPER FLOOR TYPES			
Type	**What is it?**	**What are the advantages?**	**What are the disadvantages?**
Timber joist	Timber joists span between walls, plasterboard is fixed to the bottom and timber floorboards to the top. Soundproofing regulations will require some additions to this basic scheme (see pages 147–149). Relatively thermally lightweight, although fairly thick and heavy tiles can be laid on a structurally upgraded floor.	• Dry and fast to construct. • Made of light, easy-to-handle components that are easily available. • Adaptable to changes, last-minute decisions and corrections. • Service penetrations are easy. • Environmentally friendly, especially if timber from a renewable source and not imported too far. • As above except that components have to be pre-ordered.	• Prone to squeaking unless built from well-selected timbers and well constructed. • May be damaged if put down early on in the build process; temporary decking can be used to provide a working platform, or the floor can be built after the walls. • Insulation is required between the joists for soundproofing.
Engineered timber joist	As for a timber-joist floor but the joists are beams that are I-shaped in profile, consisting of timber battens top and bottom and either a thin board or light steel struts between them to form the 'web' of the beam.	• Straighter, lighter and more dimensionally accurate than solid timber. • Quite large pipes and ducts can be cut through the web of the beam, or pass between the metal struts.	• The same as a timber joist floor, except much less likely to squeak and capable of long spans. • More expensive. • Come as a system with their own detailing and fixings – tradesmen may not be familiar with them.

Type	What is it?	What are the advantages?	What are the disadvantages?
Beam and concrete block	Concrete beams shaped like an inverted 'T' with concrete blocks (sometimes referred to as 'pots') resting between them. A concrete screed or topping is required for soundproofing. A lining or coating must be applied to the underside for soundproofing. The supplier will work out a layout and a schedule of suitable beams and blocks. They will supply only, or supply-and-fix.	• Provides a durable working platform. • Relatively environmentally friendly (especially if Aircrete infill blocks are used). • Service penetrations through the floor are easy to achieve by leaving a block out. • Fast and dry (less so if a screed is applied).	• Beams need to be delivered close to their final position by the delivery truck or a small crane, as they are very heavy to lift. • Corners and edges can be tricky, so a supply-and-fit contract with the manufacturer is recommended.
Hollowcore	These are concrete slabs, generally 1200 mm wide and 150 mm thick, with perforations through their length. They span between supporting walls in the same way as beam and block, and take the same insulation and thicknesses. They are cambered and so may require a levelling screed.	• Fast to install. • Provides a durable working platform. • Provides maximum structural efficiency. • Can be provided with long spans in excess of 16 m. • Dry (less so if a screed is applied). • Very soundproof.	• A crane close to the building is required. • Small holes can be drilled but large holes have to be cast into the plank in the factory, so you would need to plan your services carefully in advance. • Lead-in time for ordering. • Very difficult to adapt to changes, last-minute decisions and corrections. • Difficult to adapt to non-rectangular shapes.

UPPER FLOOR TYPES (continued)			
Type	**What is it?**	**What are the advantages?**	**What are the disadvantages?**
In-situ suspended slab	A reinforced concrete slab, resting on load-bearing walls, cast on temporary support.	• Provides a durable working platform. • No limitation on shape so good for odd-shaped buildings. • Very soundproof.	• Engineer required to design it. • Need to get the concrete wagon close to the building (pumping is expensive). • Need to plan your services carefully in advance. • Uses a lot of concrete and reinforcement. • Very difficult to adapt to changes, last-minute decisions and corrections. • Slow – support has to be erected and left until the floor has cured. • This kind of floor has to be designed by a professional engineer and installed by a main contractor or specialist subcontractor.

PITCHED ROOF FINISHES

Type	What is it?	What are the advantages?	What are the disadvantages?
Clay tile	Tiles come in a variety of sizes and styles, from small, flat plain tiles to large complex interlocking tiles. Tiles are hung on, and as necessary, nailed to, timber battens over a waterproof underlay. Ridges, hips and valleys are made watertight with special accessories such as ridge tiles and valley gutters. Some interlocking tiles allow a pitch as shallow as 12 degrees.	• Wide choice of styles. • Durable. • Fairly adaptable to complex shapes (less so for large tiles). • Reusable.	• Minimum pitch (depending on tile).
Concrete tile	Tiles come in a variety of sizes and styles, from small, flat plain tiles to large complex interlocking tiles. Tiles are hung on, and as necessary, nailed to, timber battens over a waterproof underlay. Ridges, hips and valleys are made watertight with special accessories such as ridge tiles and valley gutters. Some interlocking tiles allow a pitch as shallow as 12 degrees.	• As clay tile, but cheaper.	• As above, but loses appearance with age.

PITCHED ROOF FINISHES (continued)			
Type	**What is it?**	**What are the advantages?**	**What are the disadvantages?**
Stone	Various types of 'sedimentary' stone cut into thin sheets, nailed to battens over a waterproof underlay on steep pitched roofs. Clay accessories or metal flashings are used to form ridges, hips and valleys.	• Attractive. • Durable. • Fairly adaptable to complex shapes if stone easy to shape. • Reusable.	• Heavy so requires a more substantial roof structure. • Laying is a difficult, specialist job. • Cannot make curves, except with small units. • Roof pitches tend to be steep (around 45 degrees) as the rough stones do not fit very closely together.
Slate	A particular kind of stone that splits into thin sheets, is cut into rectangles and nailed to battens over a waterproof underlay. Ridges, hips and valleys can be finished with metal flashings or clay accessories. Suitable for roofs with a pitch of at least 30 degrees.	• Light. • Durable if good quality. • Low environmental impact if sourced from UK. • Reusable.	• For curves it's necessary to employ a skilled slate roofer.
Glass	Glass, usually double- or triple-glazed units in framing system. Can be mounted in a proprietary 'capping' (fixing) system, on timber rafters, or in a proprietary metal framing system such as 'patent glazing'.	• Beautiful effects possible.	• Relatively poor insulation. • May cause excessive summer heat gain. • Requires cleaning, which may be difficult. • Difficult to 'black out' if darkness wanted.

PITCHED ROOF FINISHES (continued)

Type	What is it?	What are the advantages?	What are the disadvantages?
Lead	Metal sheets fixed together with various types of folded and lapped joints or seams. Superficially similar, but each metal has a different system of joints and details. Laid on boarding (usually plywood) or direct to rigid insulation via a clipping system. Lead is grey but has more character than the 'flatter' greys of steel and zinc.	• Sheets can be applied to any pitch over 5 degrees. • Light, very durable and can be formed to awkward shapes, integral gutters and curves. • Very durable. • Recyclable.	• Some concerns about toxicity of run-off. If you have any concerns please contact the advisory organization.
Copper	Weathers to brown, green or even blue, depending on air quality and exposure.	• As above.	• As above. • Environment concerns about pollution from copper mining.
Stainless steel	Striking bright appearance not to all tastes but can be supplied 'pre-weathered' grey (to various shades from pale to almost black).	• As above. • Very environmentally friendly: stainless steel is a recycled product and is recyclable.	• Tends not to lie flat, giving rather 'industrial' appearance.
Zinc	Available in a range of shades.	• As above.	• Requires fleece ventilating layer under the metal sheet.

PITCHED ROOF FINISHES (continued)

Type	What is it?	What are the advantages?	What are the disadvantages?
Thatch	Bundles of reeds or straw-laid courses to a thickness of about 300 mm, fixed down to the roof structure by timber or metal rods, in turn fixed down by long thatching nails. Fire-resisting boards are normally required to be laid under new thatch. The ridge is finished by wrapping thatch over. No accessories are used for hips and valleys — the thatch is just 'swept' round. Thatch should be laid to a pitch of at least 45 degrees (50 degrees is preferred).	• Light. • Beautiful effects possible. • Adaptable to awkward roof shapes. • Minimal environmental impact in production, use and disposal.	• Some care needed to prevent fire: although chances are no greater than with conventional roofs; if a fire does take hold the results will be more damaging than on more conventional roofs. • Usually thatched roofs must be 12 m away from site boundaries owing to regulations to control fire spread, although the rule is often relaxed if appropriate fire measures are taken.
Timber shingles and shakes	Thin tiles of timber (usually a durable species like oak, chestnut or western red cedar). Shingles are sawn, shakes are split and so have a ridged surface. Fixed down to timber battens over an underlay.	• Light. • Attractive. • Minimal environmental impact in production, use and disposal (if certified).	• Run-off from western red cedar corrodes some metals (such as lead flashings) after prolonged contact. • Once or twice a year, dirt build-up between some tiles needs to be removed with a wire brush, especially if the roof is close to trees.

FLAT ROOF FINISHES			
Type	**What is it?**	**What are the advantages?**	**What are the disadvantages?**
Mastic asphalt	A mixture of bitumen and limestone applied molten to most roof structures to create a durable, waterproof membrane.	• Can be applied to awkward shapes. • Resistant to impact damage. • Covers up any unevenness.	• Highly dependent on quality of design and installation for durability. • Bitumen fumes given off during installation (health fears but no proven risk).
Three-ply polyester or built-up roofing	Formerly known as felt, now made of synthetics. Three layers of sheet material stuck down with hot modified bitumen. The top sheet usually has a reflective mineral coating. There are several specifications, the best being the most durable. Applied to decking, usually plywood, over a ventilated air space.	• Quick. • Attractive, with a choice of colours.	• Highly dependent on quality of design and installation for durability. • Replacement includes the substrate (timber-based board or rigid insulation) to which the roofing is bonded.
Single-ply	Very thin (as little as 1.2 mm) synthetic/rubber-based sheet, applied to decking, usually plywood, or to rigid insulation board.	• Easy to repair by patching (although this looks unsightly). • Can be applied to awkward shapes. • Becoming common (around 10 per cent of the market). • Rubber-based varieties are quite environmentally friendly.	• Highly dependent on quality of design and installation for durability. • Very susceptible to impact damage (easy to puncture). • Can look untidy unless specified as being visible.
Lead, copper, stainless steel, zinc	As described in the table of pitched roofs.	• As described in the table of pitched roofs.	• As described in the table of pitched roofs.

FLAT ROOF FINISHES (continued)

Type	What is it?	What are the advantages?	What are the disadvantages?
Glass	Glass, usually double- or triple-glazed units in framing system. Can be mounted in a proprietary 'capping' (fixing) system, on timber rafters, or in a proprietary metal framing system such as 'patent glazing'.	• Beautiful effects possible.	• Relatively poor insulation. • May cause excessive summer heat gain. • Requires cleaning, which may be difficult. • Difficult to 'black out' if darkness wanted.
Liquid waterproofing system (LWS)	Chemical compounds that can be applied to an existing roof in liquid form and which quickly dry to form a solid waterproof membrane just a few millimetres thick. Most can be installed cold, typically by brush, spray or roller.	• Can be applied to a wide range of otherwise unsatisfactory substrates. • Easy maintenance by re-coating. • Choice of bright colours.	• Expensive over rough surfaces (as liquid fills all gaps). • Cannot be applied in all weather conditions. • Solvent fumes given off during installation.
Intensive green roof	A thick layer of soil that can support a wide range of planting, on a single-ply membrane, drainage layer and root barrier on fairly shallow pitches (soil tends to 'creep' down the roof).	• Beautiful effects possible. • Provides wildlife habitat. • Reduces peak rainwater flow to drains.	• Very heavy (especially when wet). • May need irrigation, or might look unattractive in very dry periods. • Will need maintenance (especially to weed out species with invasive roots).
Extensive green roof	A thin layer of soil supporting a limited range of drought-resistant alpines such as sedum, on a single-ply membrane, drainage layer and root barrier.	• As above, but lighter. • Steeper pitches possible.	• As above but less effect on rainwater loss and less species-rich.

INSULATION TYPES

Type	What is it?	Where is it used?	What are the advantages?	What are the disadvantages?
FIBROUS				
Mineral fibre/glass fibre 'quilt'	Rock or glass spun at high temperature into fibres and supplied as a rolled-up soft slab.	• Attic floor, timber-frame walls and floors.	• Easy to cut and fit. • Can be installed compressed so that it expands to fill gaps.	• In the past the coarse texture used to irritate the skin. Technology has now moved on and it's softer and easier to handle. It can be obtained pre-wrapped.
Mineral fibre/glass fibre batts	As above, but supplied as a semi-rigid board.	• As above, and masonry wall cavities.	• As above	• As above
Cellulose fibre (loose)	Recycled newsprint processed and supplied packed into bales. Sprayed into place using special spraying equipment.	• Attics, timber-frame roofs walls and floors. • In 'EVT' or 'breathing' constructions in which vapour can diffuse naturally.	• Particularly good at filling awkward shapes and gaps. • Helps keep timber frames dry. • Environmentally friendly.	• Must be installed by specialist. • Must be kept dry. • Installation can be messy (dust).
Cellulose fibre batts	The same material as above but supplied as slabs.	• As above.	• Can be installed compressed so that it expands to fill gaps.	• The slabs can be fragile and difficult to cut or handle without breaking up.
Sheep's wool batts	Sheep's wool processed into a slab.	• As above.	• Can be installed compressed so that it expands to fill gaps. • Probably the most environmentally friendly if using British wool. • Easy, pleasant, safe and healthy to handle/use.	• Sheep's wool is not as effective as other fibrous insulants. However, greater insulation properties can be achieved by using more wool to a greater thickness.

INSULATION TYPES (continued)

Type	What is it?	Where is it used?	What are the advantages?	What are the disadvantages?
RIGID BOARDS				
Expanded polystyrene (EPS)	Cellular plastics material made by foaming polystyrene into little white balls that are then formed into a slab.	• In masonry wall cavities, concrete floors, pitched and flat roofs.	• Easily available. • Strong.	• Messy when cutting. • Less easy to fit around awkward spaces than other products.
Extruded foamed polystyrene **Polyisocyanurate foam (PIR)** **Polyurethane foam (PUR)** **Phenolic foam**	Foamed plastics material cast into a slab.	• As above and in timber-frame walls and floors.	• Versatile. Some types come with chipboard and can be walked on (ideal for storage spaces in lofts). • Ideal for areas where space is restricted; the right insulation properties can be achieved while taking up less space.	• Although new products no longer contain ozone-depleting chemicals, they may still contain pollutants – check with the manufacturer if you have concerns.
Cork	Bark of the cork tree compressed into boards.		• Very environmentally friendly.	• Less powerful. You need to use more of the material.
Foamed glass	Created from molten glass and carbon and baked to form a rigid slab.	• As above, and where very high compressive strength is needed.	• Particularly strong. • Waterproof. • Easy to cut.	• Less powerful. You need to use more of the material.

INSULATION TYPES (Continued)

Type	What is it?	Where is it used?	What are the advantages?	What are the disadvantages?
OTHER				
Foils	This comes in thick sheets, bonded to a thin foam board or layered with foam and wadding.	• Versatile insulation	• The sheet is easy and quick to fix. • The boards are light. • Saves on space.	• There are doubts about its insulation effectiveness. • It is expensive.
Sprayed foam	Plastic foam in liquid form that forms a quick-drying foam spray. Available in cans or installed by specialist.	• As temporary remedial measure to the underside of old tied or slate roofs. In awkward areas of timber-frames.	• Can be used to fill and seal difficult areas.	• Less powerful. You need to use more of the material.

NB *Protective clothing and masks should be worn when fitting insulation in a confined space.*

6 Managing the build

Before you start on your project you need to consider who is going to manage the construction work. There are a number of options and you should be aware that all building and construction projects need managing to some extent, and that the quality of the management will have a direct impact on the success of the build.

Contractors and project management

Essentially, you have to decide whether you want to manage the build yourself or intend to delegate the process to someone else. If delegation is your preferred choice you have three options: hire a builder, enlist the services of an independent project manager or employ your architect to manage the project.

SELF MANAGEMENT

A large number of people choose this option and manage the build very successfully. Do not be put off if you do not have any building experience or if you feel lacking in inspiration. The skills you need to succeed are good organization, the ability to work with others and to be able to tackle problems confidently and calmly.

Managing the project yourself means that you can monitor every stage of the build, making sure it stays on budget and schedule.

Your job will involve employing subcontractors, ordering materials, hiring plant and generally making sure the build progresses as smoothly and efficiently as possible. All the technical problems will be the responsibility of your subcontractors or house designer. Before choosing this option, you need to consider how much time and energy you can devote to project management. Although you will not have to be on site 12 hours a day, you will need to be available and able at times to drop everything in order to help solve a problem or order more materials.

Above all, it is essential that the site runs efficiently. For example, if you are providing the materials, you need to make sure that they are available in the correct quantities when they are needed. If they are there too early they will be in the way and risk getting damaged; too late, and you may face costly delays. You might find that your subcontractors will charge you for waiting time and they may even leave your job to go and work for someone else.

Your build will become a major priority for you, affecting both your work and your home life. Despite that, the experience will change the way that you approach situations in the future as it will give you a new-found confidence to deal with problems head on. Furthermore, there is nothing to rival the satisfaction that you will gain from taking such an active role in creating your dream house.

EMPLOYING A MAIN CONTRACTOR

This is the most common route for self-builders, and involves employing just one builder. This may be a single individual or a building company. Either way, they have a contract with you to build your

SELF MANAGEMENT

Pros	Cons
Brings the greatest savings: effectively you are fulfilling the role for free.	You take all responsibility and any mistake you make will probably cost you time and money.
No additional costs for employing someone else to do the job.	It is time consuming. You will need to make it the number one priority in your life throughout the duration of the build.
You make all the decisions.	
The most satisfying option.	

Failing to get the necessary materials and equipment on site in time for each stage of the build will jeopardize your schedule.

EMPLOYING A CONTRACTOR

Pros	Cons
Less involvement: you will be making decisions on the aesthetics of the house but not on how it is going to be built.	You pay for this service, it is not free.
It is the main contractor's responsibility to order materials, arrange labour, hire plant and tools.	You are less hands-on during the build process.
An experienced individual or company will be in charge of your build.	

Weigh up each situation in advance. If you intend to carry out a good deal of the work yourself, make sure you are up to the job.

house to an agreed design and specification. You and the contractor should settle on a fixed price and fixed timescale, both of which will be detailed in the contract.

The main contractor will normally be responsible for ordering all the materials, scheduling the work and employing a number of subcontractors if he or she is not going to carry out the works him- or herself. In the simplest terms, once you have bought your land, completed your design, and been granted planning permission, you hand over your project to your main contractor and at the end of the agreed timescale he should come to you with the keys to your new home. It is not usually as simple as this, however, and there are likely to be situations in which your input is required, for example in choosing fixtures and fittings or paint colours.

Unless you have employed your builder to design your house as well as build it, you are still responsible for the design of the house (normally carried out by an architect, house designer or package company). This means that it is your responsibility to solve any problems the builder might have with the drawings. You can ask your builder to contact the designer directly with queries regarding the design – they are usually simply a question of clarification, for example dimensions missing or confusion between two drawings – but it is important that you make sure the questions are answered promptly so as not to delay the progress of your build.

Employing a single main contractor to carry out your build is a lot less stressful than carrying out the project management yourself, but this comes at a cost. Depending upon the level of their involvement you can expect this option to cost between 10 and 20 per cent more than the self-manage option. This has to be weighed up against the reduced level of input that is required by you.

EMPLOYING AN INDEPENDENT PROJECT MANAGER

A popular option is to employ an independent professional project manager to undertake the management of your build. He or she will be responsible for all aspects of the build, from finding subcontractors to developing the work schedule and ensuring that your house is completed at the right time for the right price. An independent project manager will act as the intermediary between you and the subcontractors carrying out the build. He or she will be the one dealing with all the technical queries and design issues, although those regarding the finished aesthetics of the house may still come to you.

A project manager will either charge by the hour (this can be between £30 and £60 per hour depending upon location and experience) or, more commonly, he/she will charge a flat fee based on a percentage of the build budget. Again, depending on the location, experience and the level of involvement of the project manager, this can be between 5 and 15 per cent of your build budget (it is often at the lower end of this scale). Many project managers feel that you will save this amount (and possibly more) against the cost of employing a main contractor.

Do not expect to see the project manager on site all of the time. He or she will make the necessary visits to ensure that works are progressing to the schedule. If you do decide to employ a project manager, make it a condition of his or her employment that you have regular (say fortnightly) site meetings for an update on progress and issues that need discussing. You could also ask for weekly progress reports so you can see how works are going on site.

If you and your partner both have demanding jobs and a busy family life, you should certainly consider employing a professional to run the project for you.

EMPLOYING AN INDEPENDENT PROJECT MANAGER

Pros	Cons
An independent professional will be managing the build for you – not someone who has an eye on profit level or personal design intentions.	You pay for this service – it is not free.
The build should be less involved: you make decisions regarding the aesthetics of the house and not on how it is going to be built.	The project manager will not be on site full time, which means that works are carried out unsupervised.
You should have no responsibility for ordering materials, arranging labour or hiring plant.	You are less hands-on in the build process.
An experienced individual or company will be in charge of your build.	

USING YOUR ARCHITECT OR ARCHITECTURAL TECHNOLOGIST AS THE PROJECT MANAGER

If you are using an architect or an architectural technician to design your house and you don't feel able to carry out the project management yourself, it is worth considering asking him or her to take on this role. Most architects learn about project management as part of their training and may well have carried out this role.

This option does have advantages: by employing one professional to carry out two roles you should make a saving on fees; furthermore, in theory, the architect will be fully responsible for both the design and project management of your build. This is a reasonable responsibility however, and you must feel confident and comfortable that your architect is capable of the task. It is possible for you and your architect to have different ideas as to the level of project management he or she is taking on, and it is advisable to have this outlined in writing.

WHAT IS INVOLVED IN MANAGING MY BUILD?

Very simply, managing your build involves two key activities – planning and coordination. There are many other tasks that you carry out, but without planning and coordination your project will not succeed. The secret to a successful building project is making sure that the resources (labour and materials) are on site when they need to be in order for the works to follow the schedule to completion.

Carry out any work you do yourself to the highest possible standard. Plan your time, buy or hire the proper tools and take advice from an expert if necessary.

Any problems need to be addressed and resolved quickly: you need to be a good communicator, motivator and sometimes peacemaker.

Many project managers become obsessed with the date that any particular tradesperson is due to start on. While this is important, it can make more sense to focus on the date that same tradesperson is due to finish the work. Building and construction is all about coordinating trades: one trade will need the

USING YOUR ARCHITECT OR ARCHITECTURAL TECHNOLOGIST

Pros	Cons
An experienced professional will be managing the build for you.	You pay for this service – it is not free.
The architect or architectural technologist is responsible for the design and the management of the build. This reduces the risk of design problems delaying works on site.	You may still have responsibility for some aspect of the project management, for example, ordering materials, arranging labour, hiring plant.
An experienced individual or company will be in charge of your build.	The architect or architectural technologist will not be on site full time, therefore works will be carried out unsupervised.

input of another before it can start or finish the work. Simple examples include the bricklayer who cannot start laying bricks until the groundworker has finished pouring the foundations, or the plasterer who cannot start on the rendering until the carpenters have fixed the door linings.

It may sound very simple but it is very easy to falter in situations like this and the consequences can be considerable.

MANAGING SUBCONTRACTORS

Keep in contact with them throughout the build. They need to be appointed early – ideally before you start any building work. An initial appointment will be based on the dates from your schedule of works, but this may change (for better or worse). Make a point of speaking to subcontractors regularly to give them up-to-date site progress and to inform them of any changes that affect them. As

WHO'S WHO IN THE BUILDING TRADE?

Name	What do they do?	Name of professional body
Architect	Designs all aspects of your house.	• Royal Institute of British Architects (RIBA).
Architectural technologist	Designs all aspects of your house.	• British Institute of Architectural Technologists (BIAT).
Structural engineer	Makes sure that your house is structurally sound.	• Institute of Chartered Engineers (ICE).
Quantity surveyor	Provides you with a detailed estimate of the cost of building your house, assists in appointing a main builder and deals with any contractual disputes.	• Royal Institute of Chartered Surveyors (RICS).
Project manager	Manages the construction of your home.	• Association of Project Managers. • BuildStore's Project Coordination Service. • Chartered Institute of Building. • Royal Institute of Chartered Surveyors (RICS).
Main contractor/ builder	Responsible for the build of your house.	• Federation of Master Builders. • Chartered Institute of Building.
Groundworker	Excavates and concretes the foundations, pours concrete ground-floor slab, installs underground drainage. May do hard landscaping.	• Federation of Master Builders. • Chartered Institute of Building.
Bricklayer	Lays all bricks and blocks, including the external shell of your building and internal partitions if in blockwork.	• Association of Brickwork Contractors. • Federation of Master Builders.

WHO'S WHO IN THE BUILDING TRADE? (continued)

Name	What do they do?	Name of professional body
Carpenter	Fits the roof trusses, all doors and frames, architraves and kitchen.	• The Institute of Carpenters.
Roof tiler	Felts and battens the roof and lays the roof tiles/slates.	• National Federation of Roofing Contractors.
Electrician	Installs the electrical system in your house (including all sockets and light fittings).	• National Inspection Council for Electrical Installation Contracting (NICEIC). • Electrical Contractors Association (ECA)
Plumber	Installs hot and cold water systems, heating systems (including underfloor heating), fixes all sanitary ware (baths, toilets, showers), fits gutters and downpipes.	• Heating and Ventilation Contractors Association (HVCA). • The Institute of Plumbing and Heating Engineering (IPHE). • Council for Registered Gas Installers (CORGI). • Association of Plumbing and Heating Contractors (APHC).
Ceramic tiler	Fits kitchen and bathroom wall and floor tiling.	• The Tile Association.
Plasterer	Does all plastering, rendering and plasterboarding.	• Federation of Plastering and Drywall Contractors (FPDC).
Painter & decorator	Does all painting and decorating, including any staining of timber.	• British Decorators' Association.
Window installer	Installs windows and doors.	• British Woodworkers Federation (BWF). • Glass and Glazing Federation. • Plastics Window Federation. • Steel Window Association.
Scaffolder	Hires and erects scaffolding to enable access as your build progresses.	• National Access and Scaffolding Confederation.
Hard/soft landscaper	All of the external paving, top soiling, driveway construction and turfing.	• British Association of Landscape Industries.

NB: *Most tradesmen/subcontractors charge by the day (or, minimum, half day). Normally rates include small tools that a tradesman may carry.*

you get closer to their start dates, ask them to come to the site to make sure everything will be ready for them. You can establish a good working relationship in this way. As long as subcontractors are kept informed, they should balance any other work with yours so that they are available when you need them.

Scheduling the works

Successful project management relies on good and detailed planning. Whether you are a first-time project manager or someone with 30 years' experience you will need to have a detailed schedule of works or Gantt chart. More commonly known as a 'programme', this sets out the work to be done in a graphical form, showing what needs to be done and when, and is a good tool for managing a builder's progress.

If you intend to take on the project management yourself, the schedule of works will be one of the first tasks you carry out. You may even find that your mortgage lender requires such a programme as a

THE SIX STAGES OF HOUSE BUILDING

Building stage	What happens?
Stage 1: Land purchase	
Stage 2: Foundations and drainage	Excavate foundations; concrete foundations; brickwork to damp-proof course (DPC) level; internal underground drainage; concrete ground-floor slab; backfill and cavity filling.
Stage 3: To wall-plate level (for brick and block) or timber-frame	Brick-and-block house: Brickwork and blockwork to first lift of scaffolding (see page 169); first lift; brickwork and blockwork to wall plate. For timber-frame house: arrival and erection of timber-frame.
Stage 4: Wind- and watertight	Roof trusses/roof carpentry including fascias and soffits (undersides); felt and batten; roof tiling; roof leadwork, flashings; rainwater guttering/downpipes; installation of windows and doors.
Stage 5: First-fix	External painting; dismantle scaffolding; service connections; external drainage; electrical first-fix; plumbing first-fix; heating first-fix; carpentry and joinery first-fix; plastering/plasterboard; taping and jointing.
Stage 6: Second-fix to completion	Joinery second-fix; electrical second-fix; plumbing second-fix; heating second-fix; wall tiling; painting and decorating; kitchen fitting; sanitaryware; mastic sealants; service connections; topsoiling; external works/driveway; snagging; final clean; issue of Building Control Completion Certificate; issue of structural warranty from insurer.

NB *Follow these headings and allow realistic times periods to carry out the works to produce your working programme.*

COMMON MISTAKES

- Allowing insufficient time for completing the work. This will throw your entire schedule off course. If your trades people have based their start dates on your programme, you will have a major job preventing significant delays. Be realistic with time allocation: talk to your subcontractors to find out how long it should take to complete an activity. If in doubt, allow extra time.
- Not overlapping works. It is not always necessary to have one task or one trade completed before the next one starts. For example, once your carpenter has fixed the door linings on the ground floor your plasterer or dry liner can commence working while the carpenter moves on to the first floor. Sensible overlapping of works can reduce your programme duration significantly.

condition of your mortgage offer. You can find examples of programmes in self-build books and magazines. All you need to know is the basic order in which a house is built and how long each stage takes. Use the experience and knowledge of your professionals in putting together a programme that everyone is happy with. Target a realistic completion date – challenging but achievable. (If you are using a package company to design your house they will probably produce a programme for you.)

GETTING HELP

If you feel unable to produce a programme, it is possible to employ a professional to do it for you. An independent project manager will carry this out in a relatively short period of time, for a fee of between £500 and £1,000. You will need to provide him or her with copies of your drawings, specifications and any other documents that relate to your house.

A critical aspect of project management is having a good grasp of the order in which things take place. It is essential to have the right people on site at the right time.

ANTICIPATING AND PREVENTING DELAYS

One of the biggest causes of programme (and budget) overrun in the construction industry is poor or incomplete design.

Make sure each subcontractor receives the architect's or designer's latest plans two weeks before they are due to start works on site. Ask them to review the plans and to confirm that they contain all of the information and details that they need to do the work. They should tell you where there are any shortfalls so that you can pass them on to your designer, who can resolve them before works commence on site.

If you change your mind about what you want, you should expect this to affect your programme. The extent of the impact will depend on when you make the change and how much work it involves. Some alterations are quite simple – for example if you decide to have full-height ceramic tiling to your bathroom walls instead of half height: as long as the tiling has not been started already, it is unlikely to cause a major delay. If you decide to change your windows from PVC-U to softwood a week before they are due to be fitted, this will cause a major delay to your programme.

Always speak to your subcontractor or builder to ask whether a change is possible and what sort of impact it will have on the completion date. Once you have had this discussion you can make an informed decision about whether the change is absolutely necessary given the time and cost implications of carrying it out.

Unforeseen events may cause you problems on site. These usually concern the ground, for example if you are digging foundations you may come across an unexpected live sewer or you may find that roots from nearby trees are deeper than you expected. In these situations you have to be pragmatic: use the experience of your contractors to come up with a solution. Such unforeseen circumstances will cost money, and it for this reason that it is essential to have a reasonable contingency in your budget.

COPING WITH THE WEATHER

You can control the majority of what happens on site but you cannot control everything. You should expect the weather to have an impact on the progress of your build. This is normal and most contractors will make allowances for it when considering how long it will take to do a job. If it is too cold you may not be able to pour concrete or lay bricks, too windy and you may have problems erecting roof trusses, too wet and your trenches or house may flood; too hot and you may have problems with concrete setting too quickly. The best action to take is to pay close attention to long-term weather forecasts. If poor weather is likely, discuss this with your subcontractors and decide what to do.

Dealing with problem subcontractors

It will be your responsibility to deal with any problems that arise with your builder or subcontractors. The best advice is to tackle each new situation as it arises and not to ignore it. Most issues will be trivial and minor and addressed very quickly and simply. However, there are a few core problems that you may have to face and resolve without falling out with the builder or subcontractor.

MAJOR CAUSES OF DELAY?

Whether you are project managing the build yourself, or have enlisted the services of a professional, you should be aware of the major reasons for works not progressing in accordance with your programme:

- not appointing subcontractors early enough
- subcontractors not performing
- subcontractors not turning up (because they are working elsewhere, or you are not paying them)
- materials shortages
- plant not available
- design problems
- design changes
- unforeseen items (for example, poor ground conditions)
- poor weather.

NB See pages 160, 162 and 166 respectively for issues regarding subcontractors, materials and plant.

Do not allow problems to escalate. If you think a job has been done badly or something has been overlooked, raise the issue with your contractor immediately.

POOR WORKMANSHIP

Act promptly: speak to your builder or sub-contractor, show him or her the problem, explain why you are not happy with it and what you want done. In the majority of cases this is sufficient, but if you are still not happy, write to the person and record the problem. If you feel strongly, you can withhold money against the builder or subcontractor until the work is rectified to a satisfactory standard. To avoid legal problems, you must make sure you state the reasons for withholding money in writing. Should this not resolve the problem, consider employing an independent expert to act on your behalf, for example an architect or project manager.

SLACK TIMEKEEPING

There are likely to be times when your site is quiet because your builder or subcontractor is working on another job. This is often unavoidable as nearly all tradespeople juggle projects to keep different clients happy. Try not to get too concerned at the odd day that no one is on site but if days turn into weeks, you need to speak to your tradesperson and stress your anxiety at the lack of presence on your site. Having a contract with a fixed completion date will certainly assist you in maintaining progress. It is a good idea to keep a record of who is on site and when, so that you have evidence to back up your claims, should a dispute escalate.

OVERRUNNING ON A JOB

The best way to avoid this is by having a formal contract with your builder that states both a start date and completion date for the project. This contract should include a clause that binds the builder to pay damages if he or she fails to complete on the agreed date. Damages are expressed in a set amount per day or per week. If you are living in rental accommodation during the build and it costs you £200 per week, it would not be unreasonable to include a damages amount of £250 per week. It is advisable to have an agreed programme or schedule of works for your build.

ADDITIONAL COSTS

You should have a written quote from your builder before he or she commences work. This should be detailed, stating clearly what is included within the quote and what is not. Once happy with the quote, you should make it legally binding by confirming your acceptance in writing and asking the builder to proceed with the works.

A builder should only alter his or her original quote if there has been a change in the brief. Before you fall out with your builder over additional costs, think about how they might have come about. For example, did you delay progress by not choosing a paint colour or floor tile, or did you ask him or her to move a light switch? It is easy, and understandable, to forget discussions you have had or instructions you have given. Accurate records will prove invaluable in this situation.

OVERCHARGING ON MATERIALS

Although rare, this can happen when you do not agree a fixed price, but instead pay your builder an hourly rate, with materials reimbursed at cost. If you are not happy with your builder's prices for materials ask to see an invoice from his or her supplier and agree to pay the same amount plus a small percentage (normally 5–10 per cent) for his or her profit. You may not have to pay the listed price for materials as most builders can open accounts with merchants and will receive discounts. Another way around the problem is to open trade accounts and supply the materials yourself.

CHARGING VALUE-ADDED TAX (VAT)

Current legislation dictates that all new-build housing is VAT free. This means that you can recover the VAT that you have paid out during the build (although not all of it, see page 16). HM Customs and Excise produce a useful free guide, which sets out what you can recover the VAT on and what you cannot. If you are using a VAT-registered builder the process should be very simple: on submitting an invoice to you, it should be zero-rated for VAT, that is, there should be no VAT added. If he or she does charge you VAT, do not pay it, but ask for the invoice to be resubmitted.

TOP TEN TIPS FOR AVOIDING PROBLEMS

1 **Choose your builder carefully** Talk to previous clients, visit his or her sites and make sure that will be happy with working with him or her. Don't just choose a builder because he or she is the only one available.

2 **Be the boss** Always remember that you are the boss and, without being dominant, always make sure that the builder knows he or she is working for you.

3 **Form a relationship with your builder** The better your relationship the less likely you are to have problems. Those that do occur can usually be resolved quickly.

4 **Get a fixed price** Provide a detailed specification and drawings and get a fixed price for your build.

5 **Pay promptly and regularly** If you are happy to pay your builder then do so promptly. Agree stage payments but never pay in advance.

6 **Use a contract** This reduces your risk immeasurably. Unless you are confident in completing one, employ an expert.

7 **Don't let problems escalate** Deal with each one as soon as it arises and don't compile a long list of complaints for the end of the project.

8 **Keep records** of what is said and events as they take place.

9 **Get expert advice** If you are not comfortable dealing with your builder over technical matters get advice from a professional.

10 **Enjoy it** Make your site a happy one where people enjoy working. Refreshments cost very little but will go a long way to making your build enjoyable for all.

HAVING CONTRACTS WITH SUBCONTRACTORS

You can reduce likely problems significantly by picking your builder carefully and making sure all of your requirements are clearly agreed. This is best done by having a contract between you and the builder. This can be as simple as a letter that outlines your requirements and the agreed price, which you both sign and keep a copy each.

Alternatively, you can use a formal contract, the most common of which are produced by an industry group, the Joint Contracts Tribunal (JCT). These include the JCT Building Contract for the Home Owner/Occupier, which is ideal for extensions or loft conversions, and the JCT MW98 Agreement for Minor Works, which is recommended where the cost of building works is unlikely to exceed £250,000.

If you are not used to working with JCT contracts it may be worth employing a professional to complete the documents for you. If you sign a contract on which you have made a mistake, you are bound by it.

TAKING LEGAL ACTION

Most disputes can be settled amicably, by talking them through. Should this not be possible, however, it may be necessary to bring legal action against your builder or it might be brought against you. There is a dispute resolution process to follow if you have a JCT contract with your builder, but this can be time consuming and you will incur costs that may not be recoverable should you win the case.

You are better off seeking professional help to resolve the problem. Discuss the problem with a specialist solicitor or a chartered quantity surveyor. If you do not have written contract with your builder the resolution may not be so straightforward and you are strongly advised to attempt to reach an amicable solution with your builder before resorting to independent professional help.

Buying materials

You may choose to take on the responsibility for buying all or some of the materials for your build. This really is up to you and can result in a significant saving on the final cost of your build. However, it can also be very time consuming, and planning is critical if you want to avoid delaying your build and frustrating your subcontractors.

Be aware that some of your subcontractors may work as 'labour only', with you providing the materials, while others may only work for you if they provide all the materials. It is important that both you and your subcontractors are clear about who is providing what.

WHERE DO I GET MATERIALS?

Building materials can be bought in many places, from the local DIY supermarket through to the national, regional and local builders' merchants. There are also specialist brick merchants, timber merchants, sanitaryware suppliers – the list is endless. For most supplies, the best place tends to be your local builders' merchant. The majority have a wide range of products, free or cheap delivery and the level of support that you will appreciate as a novice builder.

Before you start building visit your local branches to see what they have to offer in terms of both product and added services. What are their delivery charges, for example, and how easy is it for you to get a credit account? A cheap price is a bonus, but you also need suppliers to deliver promptly and reliably.

HOW DO I KNOW WHAT TO ORDER?

There is a very good way of getting all your material scheduling carried out professionally, accurately and free of charge. Most national and regional builders' merchants will take your drawings and, after a couple of weeks, will produce an itemized schedule of all the materials that you will need to build your home. It is normal for them to charge you an initial deposit of around £150, but this is refundable against the materials bought and is only really a safeguard against you using the schedule to buy the materials from elsewhere.

When ordering materials you need to keep in regular contact with your subcontractors. They are going to be working with the materials and are the best people to advise you when to order more. During your regular visits to the site make sure that you speak to each subcontractor before you leave to ask if they need any more materials. In most cases, you need to think at least a week in advance for most supplies.

HOW DO I OBTAIN DISCOUNTS?

When you visit your builders' merchant ask to speak to the branch manager or the sales manager. Explain that you are building a new house and that you would like to discuss the materials you need. He or she will probably ask you how much you expect to spend over the course of your project (you should calculate this in advance). A simple guide is that approximately 50 per cent of your build budget will be spent on materials.

Ask what level of discount he or she would be willing to offer you – this can be upwards of 7.5 per cent and considerably more on some items. It is likely that he or she will offer you different discount levels against different materials, where it is normal to get the maximum against high-volume low-cost items such as drainage materials or timber. Get a similar quote from at least one other merchant and review the discounts that you have been offered. Decide which one to go for and negotiate a final price: they will often do what they can to match or beat any discounts that you have obtained elsewhere. When you have completed your negotiations ask for confirmation in writing.

OPENING A CREDIT ACCOUNT

Very little is paid for in cash or by credit card at a builders' merchant. Instead, you should be able to set up a credit account. This operates in a very similar way to a personal credit card in that you buy goods and at the end of every month you receive a statement, which you then have a month to pay. This is a valuable aid in managing your cash flow – particularly if your mortgage advances are in arrears. When opening an account you will be given a monthly credit limit. This will be something between £1,000 and £5,000 per month, and it is important to try and get a credit limit that matches your projected spending at its peak.

Health and safety

Building is a dangerous occupation and has one of the highest accident rates, along with agriculture and the oil industry.

In an attempt to reduce the number of accidents the Government introduced legislation called 'Construction (Design and Management) Regulations 1994', commonly known as the CDM regulations. These place duties on all those involved on a construction project to contribute to health and safety. In the main, these regulations do not apply to a domestic client. The definition of a domestic client is 'someone who lives or will live, in the premises where the work is carried out. The premises must not relate to any trade, business or other undertaking'. So as long as this applies to you, you do not need to be concerned about complying with these regulations, which is good news as they

The responsibility of good housekeeping tends to fall to the project manager. The piles of rubble on this site are potential hazards.

do have quite an administrative burden. However, under certain circumstances, the regulations do apply. These are dependent on the length of the build (if the job takes more than 30 days) and/or if the client has entered into agreement with the 'developer' (or builder contractor) who should then agree to assume all responsibilities for CDM regulations.

Responsibility for the health and safety of your workers, and anyone visiting your site, has to lie with someone, and you must ensure that either you or your construction team holds both public and employers' liability insurance.

In short, if your build is taking more than 30 days (which the majority do) somebody has to take responsibility for CDM regulations. If you are employing a builder, ensure that he does, otherwise you will have to.

HOUSEKEEPING

Most accidents on site are trips and falls – usually because the site is untidy. A simple rule is to keep things organized. If you are employing a builder it is his or her responsibility to keep the site tidy. If you are managing the build yourself, this task often gets overlooked and it falls to you to get it done. Stress to your subcontractors that you expect them to clear up after themselves throughout the build. You must hire a skip for your site and change it

It is advisable to employ a professional scaffolding company in order to make sure that your scaffolding is erected properly.

whenever it gets full. Without a skip, rubbish will accumulate, making your site untidy and dangerous.

Prices vary depending where you are in the country, but it's certainly worth ringing round to get a competitive quote.

LADDERS AND SCAFFOLDING

There are significant dangers with scaffolding and ladders. You should employ a professional scaffolding company to erect your scaffold. Potential problems occur when other subcontractors remove handrails or move ladders to gain better access to their work and forget to put them back or put them back incorrectly. Each time you visit your site make

If your site does not already have boundary walls or fences, it is worth erecting some early on in order to keep unwanted visitors out and valuable materials in.

sure that the scaffold is intact, that the ladders are in place and are secured to the scaffold, and that all the boards are in place and undamaged.

If you are not sure yourself ask an experienced subcontractor to take a look for you or ask the scaffold contractor to come on site and inspect the scaffold him- or herself.

HOLES AND EXCAVATIONS

Left unprotected, holes and excavations can be very dangerous. Most groundworkers and builders are experienced enough to know this and will make efforts to keep such areas cordoned off to prevent people or machines falling into them. This is not always practical, however, and excavations need to be left open at times, for example during the digging of your foundations. In such cases, the groundworker is usually alone on site so this should not really present a problem. Nevertheless, it is important that your site is secured and that unauthorized or unwanted visitors cannot easily enter.

PROTECTIVE CLOTHING

You must insist that you, and any persons on your site, have the correct clothing and protective equipment at all times. This includes a proper hard hat and good safety footwear. Both can be bought at most builders' merchants and are relatively cheap.

FIRST-AID BOX

You must ensure that there is a well-stocked first-aid box on your site, either provided by your main builder or by you, and that it stays on site throughout the programme of works. Make sure that everyone working on site knows where it can be found.

WELFARE

There is more to health and safety than the prevention of accidents. You also need to consider the health of everyone working on your site, including you. Site welfare is very important and involves the basic facilities that a site needs in order to work successfully and in relative comfort. Facilities include basic temporary toilets, hot and

Site safety should not be overlooked. Insist that anyone entering the site is wearing the correct protective clothing and keep the first-aid box well stocked.

cold water, somewhere dry and warm to rest and somewhere to keep clothes dry. Having toilet facilities usually means hiring chemical toilets, which can be arranged through most plant-hire outlets, for around £25 per week plus delivery. The fee covers a weekly clean and empty, together with topping up with soap and toilet paper. All you need to do is find a convenient place for it to be located on site.

Hiring plant

If you are managing your build it is likely that you will have to hire plant (tools and equipment) at some point. Commonly used items of plant on house building include dehumidifiers, mini-excavators, props and cement mixers. (See page 168 for details on scaffolding.)

Most of your subcontractors should provide their own plant. However, you may agree with certain subcontractors to provide pieces of plant for them – particularly if several trades are going to use them over the same period. When appointing subcontractors, it is useful to make sure that both of

Talk to all of your suppliers in advance of starting the work so that you can work out what plant you are likely to need to hire and when.

WHAT TO HIRE AND WHY

Item	Why?
Scaffold tower	Quick access.
Dehumidifier	Speeds up the drying out process – particularly of screed and plaster.
Scaffold trestles	For use with scaffold boards to provide a safe working platform – usually for plastering or brickwork.
Cement mixer	Mixing mortar for your bricklayers.
Plate compactors	For compacting hardcore and aggregates prior to laying concrete or paving.
Acrow props	For supporting ceilings, walls and floors.
Floor sander	For levelling and resurfacing wooden floors.
Mini excavator	Digging trenches, site clearance, grading soil and aggregates.

you are clear about who is providing what. Wherever you live there a number of plant suppliers:

- specialist, local plant-hire companies
- national tool-hire companies, found in virtually all major towns and cities in the country
- builders' merchants.

You can normally hire all your plant needs from any of these sources so it just does depends what is available and the discount you can negotiate. If you are unsure about the equipment you need, ask. If you explain the kind of job you are doing you should be directed to the right equipment.

IS IT WORTH BUYING PLANT AND TOOLS INSTEAD OF HIRING?

Many self-builders consider buying certain items of plant and, in most cases, this depends on whether they are cheaper to buy than to hire. There are other considerations, however: how much will it cost you to buy?; will you be able to sell it at the end of the project?; do you plan to use it again in the future? Many self-builders buy tools second hand (often from other self-builders), which reduces the initial outlay. If you sell it again at the end of your project it can prove an even more economical way of sourcing plant.

Make sure you buy it from a reputable dealer and that you check its working order before handing over any money.

PLANT AVAILABILITY

Most pieces of plant and equipment are available at fairly short notice of, say, one to two days, but you cannot guarantee this. You can book the hire of plant days, or even weeks, in advance and it will cost you nothing. Discuss with your subcontractors what plant they need you to provide (you should know this from their quotes) and when they need it, and insist that you need at least seven days' notice to hire anything.

Hire companies, particularly the national or regional chains, will utilize other local branches within the company should a particular item be out of stock. Should any piece of plant still be unavailable to you, ask if your hire company can 'cross hire' it for you from another hire company. This is common practice among plant-hire companies based in the same area.

RISKS INVOLVED WITH PLANT HIRE

When hiring equipment there are some important points to remember to protect both your finances and your safety:

1 **Instructions** All companies will give you instructions on how to operate the equipment but it is unlikely that they will give you full training. Construction equipment is dangerous: if you don't know how to operate it then ask for instruction. Some equipment, such as a mini-excavator, can only be used with a licence.

2 **Insurance** It is your obligation to provide adequate insurance to cover the full replacement costs of any plant you hire. If you are self-managing your build you should have already taken out full insurance to cover your build and associated plant, equipment and materials (see page 170). Make sure anything you hire is covered both in terms of the wording of the policy (does it cover hired plant?), but also the financial limit of the policy. If in doubt check with your insurer – it is a wise investment to increase the limit of your policy for the duration of hire rather than be faced with a bill of several thousand pounds should an item get stolen or damaged beyond repair.

3 **Security** Construction plant is valuable. Look after it and make sure it is locked away at the end of every day. If it is a large piece of plant, chain it to something solid and immovable.

4 **Equipment return** Return the equipment in the condition in which it arrived on site. With some items this may not be possible as the very nature of its use will make it dirty, but returning a mixer covered in mortar will result in you paying a large bill to clean it.

GETTING A PLANT-HIRE DISCOUNT

Discounts are not uncommon, particularly at the tool-hire desk of whichever builders' merchant is supplying the materials for your build. Most hire companies will offer you a delivery and collection service, which can be very convenient and saves you putting dirty tools in your car. Some companies will charge you for this, while others will deliver free of charge. This is something to bear in mind when negotiating: you need to make sure you are comparing like with like.

Tool hire is normally quoted on a weekly basis, but you should only be charged for the period that you have it on hire. Most companies count a week as five days, but you can ask about special deals for weekend or long-term hire.

RETURNING PLANT

When you have finished using the plant, you need to do what is known in the trade as 'off-hiring'. If you personally collected the equipment, you should now return it. If it was delivered, call the hire company and arrange to have it collected. The hire should cease from the time of your telephone call but it can be a few days before the equipment is collected. In this situation, make sure you read the terms of the hire carefully because, although you may have off-hired the equipment, you may still have responsibility for the insurance of the item because it is still on your site. It is also recommended that you keep a record of the time that you off-hired the equipment, as mistakes can happen and you may be billed for longer.

Hiring cranes

There may be occasions for which you need to hire a crane – for example lifting roof trusses or installing beam-and-block flooring. If you can, try to make this the responsibility of your subcontractor (although you will have to pay extra for it). If you have to hire the crane yourself there are a large

Any company supplying a crane will need a couple of days' notice and you must be confident that you will be ready for them.

number of national and local crane-hire companies that you can use.

As a private individual, the best option is to speak to crane companies and ask them to carry out a 'contract lift' for you. This is best described as a package deal which covers all aspects of crane hire including selection, issue of appropriate notices and making sure that whatever is being lifted is being done so safely. A representative of the company will meet you on site and discuss what has to be lifted and where, then choose the most appropriate size crane for your needs. On the day of the lift the crane company will probably provide the necessary men – the driver, the banksman (who acts as a lookout for the crane driver) and a slinger (who is responsible for attaching the crane to the materials for lifting). They will also make sure that any roads or pavements that need closing are done so properly. This service is not cheap – a contract lift for moving roof trusses can cost in excess of £1,000 per day. However, it is good value when you consider the risk that is removed from you.

Cranes need to be booked early and are unlikely to be available at 48-hours' notice. The crucial and demanding task is to make sure the site is ready for the crane date. Turning a crane away will be expensive and you may not be able to get the crane back for several days or even weeks, which could severely disrupt your programmed schedule of works.

Scaffolding

Unless you are building a radically designed house or an underground one you will need to have scaffolding on your site. The easy option is to ask one of your subcontractors to provide it – in most cases this would be the bricklayer. Although the easiest option, there are some key points that you need to remember.

- You will pay more for the hire: your bricklayer will charge an additional fee on the normal hire cost.
- You have to make sure that the bricklayer does not remove the scaffold as soon as the bricklaying is finished: your carpenters may need it for fixing trusses or fascia board and your roof tiler will certainly need it.

HIRING YOUR OWN SCAFFOLDING

It is not too difficult to arrange your own hire of scaffolding. As with any subcontractor, approach a

AVOIDING HAZARDOUS SITUATIONS

Scaffold can be very dangerous. While all scaffold companies make sure that the scaffold is safe on leaving the site, problems can occur when other subcontractors try to remove or adapt the scaffold themselves. Key things to look out for are:

- **Ladders** These should be tied to the scaffold firmly but are often removed.
- **Toeboards** These are placed along the outside edge of the scaffold to prevent materials (particularly bricks) from being kicked off and falling below. These are often removed.
- **Scaffold boards** These are often removed or put back incorrectly.

number and ask for a fixed quote. To get an accurate price you need to provide each company with the following information:

- Drawings of your house. They will need elevations and floor plans in order to see how high it is and the overall dimensions.
- The number of 'scaffold lifts' that are required. 'Lifts' are simply the number of working platforms required. A bricklayer can safely lay bricks up to a height of about 1.5 m before it becomes necessary to raise (or lift) working height. A normal two-storey house will need four lifts – the first two for the bricklayers to get to wall-plate level, the third for the carpenters to fix the trusses and the fourth will be for any gable brickwork or chimney.
- The period of time that you need the scaffold on site. You pay for scaffold for the initial erection and a defined hire period (say 12 weeks). If you go over this period you will be charged a further weekly rate, which should be stated in your quote. Use your work schedule to calculate how many weeks you will need the scaffold. The initial lift usually happens in week two of the brickwork and the scaffold stays there until at least the roof tiling is complete, perhaps longer if it is needed to install the windows or paint fascia boards.

When asking for a quote make sure that the following are included:

- Temporary scaffold staircase from the ground to first floor. At a minimum ask your supplier to provide a ladder.
- As a safety requirement you will need to provide a scaffold handrail around any unprotected first-floor openings – the staircase or any galleried features.

GETTING THE TIMING RIGHT

Once the first lift of scaffold is erected, you need to keep in close communication with your bricklayer to make sure the next lift takes place at the right time to enable work to keep progressing. Most scaffolding companies are flexible and will try their best to meet your dates, but you really need to give them 2–3 days' notice when you need them back on site.

One thing to look out for is that scaffolders do not like coming to the site and finding that they are unable to raise the scaffold to the next lift because it is covered with bricks, blocks or other redundant materials. At best you may find yourself faced with a bill for the scaffolders' waiting time while the materials are removed; at worst they will leave the site and not return until the scaffold is clear. You can avoid this by insisting that your bricklayer clears the area before the scaffolders arrive.

Scaffold hire is expensive and really adds up if you run into delays. Get help from a professional so that you can plan the scaffold stage of your build with precision.

Insurance

You need to make sure that your site is insured. If you are employing a single builder to carry out your build make it his or her responsibility to arrange insurance. This will cover theft, damage and personal liability insurance. You must obtain a copy of the insurance certificate before works commence to check for items that may not be covered. For example, if you supply some of the materials yourself, will they be covered? Does it cover the plant and tools of additional subcontractors? If in doubt ask your insurance broker. In most cases it is best to supplement the builder's insurance with your own additional policy for the maximum protection.

If you are project managing the build yourself, employing subcontractors and supplying some or all of the materials, you will need to cover the project with insurance (see page 21).

Temporary power and water supplies

As project manager, it will be your responsibility to arrange power and water supplies to your site. This should be seen as your first priority on completing the land purchase. Depending on your location, and the appropriate local suppliers, these new supplies can take between 4 and 12 weeks to arrange.

During the first few weeks of the build you can usually get by with a generator for electricity, and battery-powered tools. A generator is a worthwhile investment either new or used. If you buy one, it needs to be a minimum of 14kVa (kilovolt ampere – a measure of power where one kVa typically equals approximately 0.8 kWh) to provide enough power. You can buy a generator for between £250 and £350, but make sure that it can offer both 110volt and 240volt. If well looked after you can sell it at the end of your project.

Water will be needed on site at a very early stage for simple tasks such as washing down the pavement or roads after deliveries, More importantly, your bricklayers will need water for mixing mortar. Contact the new supplies department of your local water and electricity companies. They will send you an application form, which you should complete and return with a site plan. A site survey is then carried out, after which

you will receive a written quotation for the new service. Only once payment has been made will arrangements be made to provide you with a new supply.

Dealing with your neighbours

It is almost certain that you will have to deal with your new neighbours during the course of your build. You may come across problems that they might be able to solve – for example, they may be willing to let you use their water or electricity supply in the event that you cannot get connected. Or you may need to use their land for access at any given stage of the project. Your neighbours are far more likely to be approachable if you have built a good relationship with them. The best way to achieve this is to keep them informed: let them know what is going on, when and for how long. There are a few key guidelines, which should help smooth the way.

Once you have completed the purchase of your plot, it is a very good idea to visit all your neighbours and introduce yourself. Explain your intention to self-build a new family home, and show them pictures of what the house will look like and how you are going to landscape the plot. Tell them how long you expect the project to take. A week or two before you start, visit your neighbours again to let them know that building work is about to commence. Explain what will happen over the first few weeks and tell them that, although some of the work will be noisy, you will do your best to keep disruption to a minimum. Give your neighbours a telephone number and encourage them to call you should they have any concerns.

Give your neighbours regular updates as to how progress is going and tell them what is likely to happen over the forthcoming weeks. Thank them for their patience and understanding. It is always a good idea to speak to your neighbours if something out of the ordinary is going to happen on site, and which may affect them in some way. Perhaps a crane or large delivery is going to restrict access for half a day, or a particularly noisy operation is going disturb their peace. People are more tolerant if they are told of any disruption in advance. Visit your neighbours once again when the build is complete and offer show them around your new home.

Managing the build checklist

GETTING THE TIMING RIGHT

- [ ] Appoint all subcontractors at least two months before works are due on site and confirm their appointments four weeks before your start date.

- [ ] Make all applications for temporary water and electricity supplies at least eight weeks before they are needed on site.

- [ ] Check the lead-in times for all materials that are your responsibility and double-check all quantities.

- [ ] Arrange a scaffolding contractor and check with your bricklayer when the first lift will be needed.

- [ ] Check with the scaffolders the notice they require to return to site.

- [ ] Check that any cranes are booked and confirm the day before they are due on site.

A SMOOTH-RUNNING PROJECT

- [ ] Make sure you have a groundworker, bricklayer, scaffolder, carpenter, roof tiler, plumber, electrician, plasterer, ceramic tiler and decorator.

- [ ] Make sure the builder has copies of the latest drawings.

- [ ] Make sure the builder and all subcontractors have copies of the latest schedule of works.

- [ ] Open an account with a builders' merchant.

- [ ] Confirm with the builder/subcontractors what plant and materials you are expected to provide.

COMPLETING THE PAPERWORK

- [ ] Comfirm with your designer that the planning authorities are happy for you to commence work on site.

- [ ] Confirm and agree your start date with the builder/groundwork subcontractor.

- [ ] Write to the Building Control Department to confirm your start date.

- [ ] Advise your structural warranty provider when you are going to start on site.

- [ ] Make sure the site and the build are adequately insured.

- [ ] Arrange for the building control officer to come to the site and issue a certificate on completion.

- [ ] Arrange for your structural warranty provider to inspect your completed house and issue the full warranty.

MAINTAINING GOOD RELATIONSHIPS

- [ ] Speak to your neighbours to let them know when work on site is going to commence.

- [ ] Agree stage payments with your builder and all subcontractors.

HEALTH AND SAFETY

- [ ] Make sure somebody has responsibility for CDM regulations.

- [ ] Make sure there is always a skip on site and change it when it becomes full.

- [ ] Check that there is a first-aid kit on site.

7 Home systems

In bringing your new house to life you will install the vital systems necessary to supply power, air and water. And these are just the basics: modern technology has brought us systems that look after our security, automate all our electrical appliances, and run wireless computer networks.

Automated systems need to be planned carefully when incorporating them into your house plans, particularly the more essential amenities that you will need on a daily basis. What demands will you have for hot water and which system will best match them? Will you need air conditioning? What about lighting? You could simply have central lights, but with the range of products available it's worth considering alternatives. There are so many options available to the new house builder that careful consideration needs to be made before committing cables and pipes to the plasterwork.

Your heating system

Choosing a boiler for a new home is not particularly glamorous, and many self-builders tend to leave the decision-making to their building contractor, who subsequently defers to the plumbing contractor. In making sure that you install a boiler that meets all your requirements, it is wise to do some research yourself in order to understand what types of boiler are available and what would suit your particular requirements.

NEW REGULATIONS FOR ENERGY EFFICIENCY

Government ruling on energy efficiency in domestic buildings requires that, from April 2005, all new boilers will have to meet even higher standards of efficiency. Specifically, the UK is committed to reducing carbon emissions by 12.5 per cent below 1990 levels, as a signatory to the 1997 Kyoto agreement. Part L of the Building Regulations is concerned with this. The Government wants all new boilers to achieve a Seasonal Efficiency of Domestic Boilers in the UK (SEDBUK) A or B rating, which

Combination, or 'combi', boilers are best suited to smaller properties, where there is likely to be a small demand for hot water.

effectively means they must be condensing. This means that only boilers that are a minimum 86 per cent efficient (that is, they convert 86 per cent or more fuel used into energy) will be permissible. However, industry committees are still sitting to decide what exceptions are likely to be allowed in properties where this is not viable.

THE CHOICE OF FUEL

For most new builds the choice of heating fuel will be gas or oil and this will influence your choice of boiler. Solid fuel and liquid petroleum gas (LPG) are also possibilities, but boilers are limited for these fuels. Natural gas is available for 70 per cent of the UK population and is the preferred fuel from an environmental point of view because it produces less carbon emission than other fuels. Heating oil (kerosene) is available to the majority of sites in the country, and although price fluctuates, it generally compares favourably in cost to gas.

Wood pellet boilers are a new technology gaining popularity in the UK, but they have been in use on the Continent for 20 years. The price of pellets is roughly the same as the price of oil but the boilers cost more.

REGULAR AND COMBINATION BOILERS

The first decision you need to make is whether you want a regular or a combination ('combi') model. A regular boiler heats water, which is then stored in a cylinder, while a combination boiler provides hot water directly at mains pressure.

For smaller properties, or where there is a relatively low demand for hot water, a combination boiler is a good choice, providing heating and instant hot water from one source. Water is not stored in the tank where it will lose heat, but is heated direct from the mains using a heat exchanger and supplied only when you need it. You can program central heating to come on and off just as you would with a traditional boiler.

This option may not suit a family, or several people wanting to use hot water at the same time, because the flow can be limited to just one tap or a given number of taps and you need a pressure-balanced shower mixer or a thermostatic shower. You also need good mains water pressure at your site. The benefit is that you don't need a hot water cylinder or a tank to store cold water. This offers an additional saving on space and avoids the expense of installing tanks and pipe work, and of long-term maintenance costs. Some manufacturers supply combination boilers that incorporate a storage tank, while other combinations have high-output and preheating systems, which aim to boost supply.

Some combination boilers can accept pre-heated water from a solar water heating system. It is best to check with the manufacturer.

CONDENSING BOILERS

Once you've decided on a regular or combination boiler, you then have the choice of either a condensing or non-condensing model. Condensing boilers heat water and pump it round the pipes and radiators. Cooled water returns to the boiler and is reheated. What makes condensing boilers so attractive is that they incorporate a secondary, or enlarged, heat exchanger that grabs back heat that would otherwise be lost through the flue to the outside air. This saves on fuel, money and carbon emissions.

A condensing boiler emits cool flue gases from which a condensate is removed through a plastic pipe. For this reason it is important to site the outlet where it won't be a nuisance to you or your neighbours. (Some condensing boiler manufacturers solve this problem by making an extra-long plastic flue, so that the plume from the flue can be piped a considerable distance away.)

Available for gas or oil, to be wall-mounted or free-standing, condensing models are both efficient and environmentally friendly, with many versions being well over 98 per cent efficient in the right conditions. Consumers are becoming far more aware of their energy efficiency and SAP ratings (see page 98) and the condensing boiler is increasingly the preferred choice.

BUYING THE RIGHT SIZE BOILER

It is important to make sure your chosen boiler has sufficient heating output for the size of your home. Buying too large a boiler to cover all eventualities is

---

SAP RATING FOR BOILERS

SAP is the UK Government's Standard Assessment Procedure for home-energy rating (see page 98). SAP ratings allow comparisons of energy efficiency to be made, and can show the likely effect of improvements to a dwelling in terms of energy use. Building Regulations require a SAP assessment to be carried out for all new dwellings and conversions. This is based on running costs for space- and water-heating, which are calculated taking account of the shape and fabric of the building, its thermal insulation, the fuel used, and the performance of the heating system. SAP ratings are expressed by a range of 1 to 120, the higher the better. SAP also delivers a carbon index, in the range 0.0 to 10.0, to indicate carbon emissions. Using energy ratings, designers, developers, house builders and home owners can take energy efficiency factors into consideration both when designing new dwellings and refurbishing existing ones.

counter-productive: there will be long periods when, even though the burner isn't firing, the boiler itself will be radiating and convecting heat unnecessarily. Opt, instead, for a modulating boiler, which can cater for occasions when you want little heat output but still plenty of hot water. If you do have an exceptional demand for hot water you could consider adding an unvented hot-water storage cylinder to your system, also known as a megaflow. This draws the water supply and heats it direct from the mains, so you need to be sure that your mains pressure meets the specification of whatever model you choose.

SITING AND CARING FOR YOUR BOILER

For the best and most efficient operation, plan to install your boiler near to the primary hot-water consumers, such as the washing machine or dishwasher, and not too far from the family bathroom. It's also vital to prevent hard-water scaling and furring pipes in your system. If you don't have a water softener, check with your installer whether you should be adding a liquid anti-scaler to the system. Many boiler manufacturers will offer advice on installing particular preventive measures at the same time as the boiler. Insulating pipes is a must.

INSTALLING RADIATORS

Unless you have opted for underfloor heating, you will need radiators in most rooms. How do you decide on size and quantity? A qualified heating engineer, such as one registered with the Institute of Domestic Heating and Environmental Engineers, will design an appropriate system for the individual case. The most common radiators are made of cast iron, steel or aluminium, although the choice of style is now vast. Cast iron is heavy, and a good traditional choice for the renovation of an older property. Cast-iron radiators are typically dark grey, however, you can paint them yourself or get them sprayed professionally. They heat up more slowly than other materials, but retain the heat for longer. Steel radiators are cheaper than cast iron and come in a wide range of sizes, shapes, colours and paint finishes. They are efficient in output – the wider the tube, the bigger the output – and are wide-ranging in design.

Ventilating your home

In order to meet Building Regulations for energy efficiency (see pages 96–100), houses have to be incredibly well insulated. While this is good for keeping the house warm and heating bills low, it is not so good for circulation of fresh air. The more efficiently you seal your home, the more the air inside it becomes stale, static and moisture-laden. Pets, cooking smells and dust mites (which thrive in humid conditions) all contribute to polluting the air. According to the Building Research Establishment Report (1996), air inside a building can be ten times more polluted than that on the street.

For some it may be simple enough to open a window to bring in fresh air. But this may also let pollution and pollen into the home. An alternative is to install a ventilation system, ideally one that combines air conditioning with air purification. The air conditioning unit's purpose is to control heating,

COMPARATIVE FUEL COSTS
Average annual cost of space and water heating for a three-bedroom home in the south east of England:

Fuel	Conventional boiler	Condensing boiler
Oil	£500	£416
Solid fuel	£588	
Natural gas	£474	£400

Latest figures from Sutherland Associates (2003), independent analysts who have been publishing comparative heating costs charts for over 20 years for the UK and Republic of Ireland.

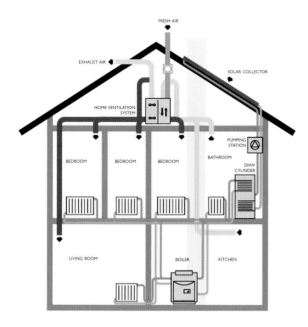

With a ventilation unit, fresh air is drawn in from outside (green) and is circulated to the living areas (red). Meanwhile 'wet' air is returned to the ventilation unit in the 'wet' rooms (yellow) and expelled.

cooling and humidity, while an air purification unit will filter the air to extract irritants such as dust, cigarette smoke, pollen and carbon from vehicle exhausts.

HOW DOES IT WORK?

A ventilation unit is usually installed in the roof space of the home and circulates fresh air from outside to bedrooms and living rooms via ductwork. Meanwhile stale air is drawn in from the 'wet' rooms in the house – the kitchen and bathrooms – and returned to the ventilation unit. The unit contains a filter that removes up to 85 per cent of pollutants, and reduces humidity to a comfortable and healthy 40–45 per cent – dust mites cannot live at this level – before extracting the air to the outside.

Some units also contain a heat extractor, in which case the system is referred to as heat recovery ventilation. This type of system recovers up to 70 per cent of the heat from extracted air as well as solar heat from the loft space. Additional solar heat, gathered through slates or tiles on south-facing roofs, can also be fed back into the fresh circulated air. In summer, you can switch to a cooling unit which links to the same system.

ARE THESE SYSTEMS VERY NOISY?

Manufacturers claim that their systems are quiet, but audibility levels will be relative to where your home is sited and any other background noise you experience. For example, if you live in the countryside you are more likely to hear the noise of the motor than you would in town. Many self-builders choose to put the ventilation unit in a garage or somewhere away from the main living area to ensure against any potential sound disruption, no matter how minor. These systems are constantly being improved, and sound levels are definitely decreasing.

IS THIS OPTION COST-EFFECTIVE?

Ventilation systems are low wattage and running costs are quoted at as little as 1p a day. However, the costs of units, ducting and installation can be around £2,000, so it may be up to 15 years before your outlay is recouped, even taking energy savings on your heating bill into account.

Central vacuuming systems

A central vacuum system turns your whole house into a cleaner. A motorized fan in the basement, or outside the house in the garage, creates suction through a series of interconnected 50 mm pipes in the walls. Central vacuum systems were made possible in the 1960s with the invention of PVC pipes, but only became more popular in the 1990s with better technology.

HOW DOES IT WORK?

Once installed, you simply attach a length of hose to one of the sockets in your home and the suction is activated. The hose length is usually about 10 m but you can have 15 m or even longer if you wish.

An average three-bedroom house will need about five sockets. Behind each socket, 50-mm plastic pipe work is hidden away in ducting within the wall cavity, and this all connects up to the power unit. The power unit uses a cyclone effect to filter dust and dirt from the air. Most manufacturers recommend emptying the unit two or three times a year, which will involve either emptying the whole unit or removing a sturdy paper bag within it.

THE BENEFITS OF INSTALLING A CENTRAL VACUUMING SYSTEM

- The systems are quieter than a conventional vacuuming system because the motor is sited away from the main rooms of the house, in the utility room, garage or basement.
- They have more powerful suction – up to five times more than a conventional vacuum.
- There is no need to drag a heavy machine around your home.
- They don't emit dust particles as you gather up debris – an advantage for asthma and hay fever sufferers.

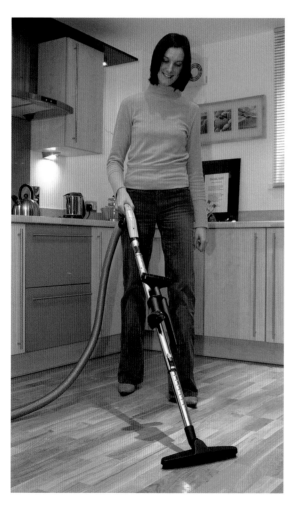

A central-vacuuming system has obvious benefits for the larger home, not least eliminating the need to carry a vacuum cleaner from room to room.

WHAT HAPPENS IF IT GETS BLOCKED?

A 90-degree bend immediately behind each socket helps to avoid blockages. If they do happen it's generally in this first section and so are not difficult to clear. First check that the dust-collecting canister isn't full. Then unscrew the end of each hose to try and locate in which of your inlets the problem lies. Manufacturers are generally helpful with advice down the phone, but if you have to call out an engineer and they can't locate the source of the blockage by using another machine to suck it out, it can be expensive and disruptive to get behind the skirting to locate the problem.

IS IT TERRIBLY EXPENSIVE?

A central vacuuming system for a three-bedroom house costs about the same as a top-of-the-range, free-standing cleaner. You can buy a DIY kit to install your own central vacuum system, saving about £200 on the fully installed version. It can be installed in a new build or conversion. If it's in a conversion, it's a matter of drilling holes in the skirting for the 50 mm pipes and threading them through and it doesn't necessarily entail pulling up floorboards or major disruption.

Home automation

The same technology used to operate your heating thermostat can be applied to a wide range of systems, which means that closed-circuit television (CCTV), home cinema, music systems, lighting and heating systems can all be controlled specifically to suit your changing needs and moods.

You can pipe music to any room in the house, set the lighting to suit your mood when you return home from work, or heat select parts of the house depending on who is at home and the time of year. Aspects of this technology are accessible from your mobile phone while you're out, or from a panel of buttons when you're in.

A good example of automation is that of home entertainment. It is normal to progress from a hi-fi to a more sophisticated multiple-speaker system, which then also incorporates the television, video and DVD. This can be customized so that televisions are not just in the living rooms, but in the bedrooms too. Once at this level, it makes sense to consider a more intelligent system that allows you to dispense with cumbersome multiple units around the house, such as several DVD players, and this can be achieved by linking everything through a central computer. The computer stores all your DVDs and CDs and this data can be accessed from all around the house.

A further optional step is to dispense with wires by using systems that converse via radio waves instead. When installing a wireless network, it is important to determine whether there will be sufficient signal strength throughout the property. Wireless networks were not originally designed for home use and elements such as the thickness and construction of walls can have an adverse effect on their performance. Instead, consider using powerline networking, which sends signals using standard mains electrical wiring. It is reliable, simple to install and can be faster than the most common wireless technology, although more sophisticated networks are being produced all the time.

CANDIDATES FOR HOME AUTOMATION
- home-entertainment system
- lighting system
- heating system
- home security system
- entrance gates
- home-computer system.

REMOTE WORKING
Home automation can be applied remotely, which means that you can programme all your systems to work normally while you are away. This brings obvious benefits to the security of your home, but also allows you to record favourite programmes and keep the house partially heated in colder months.

INSTALLING HOME AUTOMATION
If you want to incorporate home automation into your new build or refurbishment, you must get it done by a professional. Part P of the UK Building Regulations, which came into force in January 2005, makes it an offence for unqualified people to carry out electrical wiring work without having it tested and certificated to comply with BS7671, the national safety standard for electrical installation.

Naturally, it is easier to incorporate home automation into a new build, as the cabling can be integrated into the structure of the building from the

CASE STUDY: THAMES-SIDE PENTHOUSE
A two-storey luxury apartment that incorporates lighting, heating and cooling, 20 zones of electrically controlled blinds, access control interfaced with the phone system, eight audio zones, fully distributed video and television (including four security cameras) and fully distributed telephone and data. Everything is accessible by remote call, meaning it can be activated through a separate mobile phone network, as well as the main phone line.

The system was installed during the initial build, so all the wiring is hidden away. The owners simply interact with a handful of keypads and some flush-mounted touch screen panels in order to access almost anything, anywhere in the house.

CASE STUDY: 18TH-CENTURY KENT TOLLHOUSE
An 18th-century tollhouse that has had almost £200,000 worth of automated systems fitted during a renovation. The living/dining room turns into a two-level home cinema, with a plasma screen and an enormous roll-down screen for video projections. Behind a series of oak panels are racks of electronic equipment including a CD player programmed with 400 albums, a giant hard drive and amplifiers. Individual members of the family can choose to play music from their own personalized lists from anywhere in the house. The room's lighting can simulate sunsets and moonlight, and can produce many other colours.

Games can be played on any of the screens, while the sofas and beds are fitted with shakers rigged up to the sound system to add vibration while watching action scenes or playing a game. Even the garden is wired for entertainment. Weatherproof keypads control an enormous subwoofer disguised as a rock, lighting in the trees, and Internet access.

start. It is still possible in an exisiting building, but it does mean a greater level of disruption, a longer time period and more mess. Channels have to be chased into the walls, which are then plastered over, and floorboards need to be taken up.

Most installers tend to use the Cat 5 cabling found in modern office buildings, which is suitable for all home automation systems. It is probably best to install the wiring in every room of the house, with the cables returning to a central point, or wiring closet, where all the technology is installed away neatly. A primary factor to consider is that of future proofing. At a later date you might want to turn a child's bedroom into a workroom, and it would be great to have all the necessary cables installed already.

Soundproofing your home

There are two main origins of sound: impact sound, such as footsteps or light switches, which originate as impacts; and airborne sound, such as conversation, which originates as little pressure waves in the air.

Impact sound sets up little vibrations in the wall or floor. This in turn vibrates the air on the other side, causing sound in the air.

Airborne sound hits the walls and floors and sets up vibrations in them, and this gets through in the same way.

Flanking transmission occurs when you think you have created a soundproof wall or floor, but the sound finds ways around its edges, because these junctions tend to contain gaps or bits of structure that 'bridge' the soundproofing. Airborne sound gets through gaps.

WHICH ROOMS NEED TO BE SOUNDPROOFED?

The last few years have seen increased legislation relating to soundproofing – since 1 July 2003, Part E of the Building Regulations has included requirements for sound insulation between the main rooms within a dwelling. This includes the walls between a room containing a toilet and other rooms, and the walls between a bedroom and other rooms. Walls with doors and walls in en-suite facilities are excluded. However, it is sensible to build in good soundproofing throughout the house.

HOW MUCH SOUNDPROOFING DO I NEED?

Different types of material for floors or walls perform better or worse at different sound 'frequencies'. Conversation generates a lot of high frequencies, while much music contains a lot of bass, or low, frequencies. As it is complicated to express soundproofing as a range of figures for different frequencies, a single figure is derived mathematically. This is not a straightforward average, but a 'weighted' figure, designed to match the range of frequencies that tend to originate in a house. This figure is denoted by 'Rw' and its units are decibels (dB). Under current regulations, internal floors have to achieve Rw40dB reduction of airborne sound such as music or conversation.

HOW DO I SPECIFY A SOUNDPROOF WALL OR FLOOR?

House builders use standard ways of making walls and floors, many of which have been tested. Those that pass Building Regulations are described in Approved Document E. Manufacturers, meanwhile, develop new products, invest in testing them, and publish the sound-resistance of wall and floor systems that incorporate their products.

The only way to tell how soundproof these are is to test them across the range of frequencies and work out their 'Rw'. This is expensive as it can only be done by approved organizations under strict conditions.

This gives you a limited range of options. It also means that if you mix and match products in a wall or floor structure, and this results in a structure that has not been tested, there is no evidence that it complies with Building Regulations: you may be asked to test the finished wall or floor at your own expense.

WHICH WALLS AND FLOORS ARE THE MOST SOUNDPROOF?

Approved Document E ranks all the standard wall and floor types. It indicates that hollowcore is best, then beam and block, then timber joist . All of these can be improved by the addition of resilient coverings, more mass, absorbent materials, and separated ceilings. This means that, within reason, you can choose any of the standard types, and improve on it to suit your needs. Alternatively there is a range of dedicated soundproofing products, lining boards and insulation that can be used.

GETTING THE SOUNDPROOFING RIGHT FROM THE START

Soundproofing issues have been described here in very general terms, and it is a complex subject. Even standard or tested structures can fail in baffling ways, and departing from them can have unexpected results. Remedial measures can be costly and may have little or no effect if the initial problem is not understood properly.

Think about soundproofing early on in the design of your new build, as acoustic systems take up space. Otherwise, for example, linings added to walls could collide with doors. Also, structural engineers have to design to accommodate heavier walls and floors; and builders need to know what is expected of them when they price work.

A soundproof floor or wall is only as good as its weakest point, and its weakest point might be a tiny crack or a 'bridge' across an isolating layer. If air can get through, so can sound. This is why plastered blockwork is more reliable than dry lined, and why beam-and-block floors have to have a screed; both are liquid coatings that get into all the gaps and cracks before drying. Remember that concrete products and mortar shrink as they dry out after construction, and that timber-framed structures need sealant joints, because they also dry out and shrink.

The weakest point in any wall is the door which can be made more soundproof, or special soundproof doors can be used. Bear in mind that some fire-alarm systems rely on being heard through doors, and some ventilation systems rely on air leakage under doors.

Floors can also generate noise. There are many good reasons to use natural products, but timbers are seldom perfectly straight and tend to shrink, which can lead to squeaking. This should be avoidable if the floor decking is screwed down with closely spaced fixings, but a more robust solution might be to use engineered timber joists (see page 135).

For particularly noisy rooms or sensitive ears, you could look at the acoustic systems normally used in recording studios. These include heavy, sealed, fabric-covered doors, and hollow wall panels filled with absorbent material.

FIVE WAYS TO IMPROVE YOUR SOUNDPROOFING

1 **Resilient layers** Carpet reduces impact sound. If you want an exposed timber or tiled floor, there are proprietary products that provide a chipboard deck to which timber or tiles can be fixed, thus 'isolating' them from the structure below. There are many systems to choose from and they can be applied to walls as well as floors.

2 **More mass** Mass dissipates sound. You can use thicker or higher-density concrete floors or blockwork walls, or thicker or heavier lining boards.

3 **Isolated ceilings** Any breaks in the path of sound are good, and there are ways of 'isolating' the ceiling from the floor above. These range from proprietary flexible metal strips from which the lining board 'hangs' to building a completely separate ceiling structure.

4 **Preventing flanking sound transmission** Approved Document E refers you to advice published by the Building Research Establishment (BRE) on how to avoid flanking transmission of sound. Manufacturers will also advise on how to achieve this with their systems. For example, the cavity in the external wall might have a barrier.

5 **Absorption** Soaking up sound in a room reduces its passing through to other rooms. Carpet is very good for this, as are soft furnishings and fabrics.

An effective way to create more mass, and so improve your soundproofing, is to install an acoustic underlay before laying the final floor finish.

Making your home secure

All houses can have a home-security system installed after construction, but it is likely to involve raising a few floor surfaces. If you are building from scratch, it is by far the better option to install security as part of the initial build.

EFFECTIVE LANDSCAPING

One of the simplest ways to make your new home more secure is through considered landscaping and boundaries. For example, thorny or spiky plants around the edge of a garden act as a good deterrent, while non-prickly plants and badly positioned trees can prevent a break-in being spotted by passers by.

The same is true of walls and fences – they should not obscure doors and windows at the front of the property and should be as high as possible at the sides and the rear. This means that any intruder trying to break in at the front of the property would be spotted and that his or her chances of getting to the back of the house are significantly reduced.

USING LOCKS AND BOLTS

High-quality locks are essential to any security system – especially if you cannot afford an alarm. In terms of budget, it is possible to secure a four-bedroom home with high-quality locks and bolts for as little as £500. All the main doors should be fitted with a five- or seven-lever mortise deadlock. This kind of lock is operated using a key, which activates a bolt from the door into the doorframe on locking the door. The more levers, the more difficult the lock is to pick. Many insurers insist on a minimum of five levers for front doors.

For side and rear doors, which are not used so frequently, it is best to add key-operated mortise bolts to the top and bottom of the door. Some specialist lock makers produce mechanisms that strengthen the hinges on the door to prevent them being knocked down. These are advisable for maximum security. Windows also need good locks. There are many types available for either PVC-U, wooden or aluminium frames. Essentially, most types lock the window to its frame. They are usually simple to fit and inexpensive to buy.

INSTALLING STRONG DOORS

If you are renovating a property, make sure you replace any very old or weak doors or windows. If you opt for new wooden doors it is worth making sure they are as strong as possible. For example, they should be made of solid wood, no less than 44 mm thick. The door stile should be at least 119 mm wide to allow for a mortise lock. Look out for new doors and windows with built-in security measures, such as multipoint locking devices, anti-lift devices, eye viewers and security chains.

BELL-ONLY ALARMS

It is worth considering incorporating an alarm system – if only a simple one. The cheapest kind is known as a 'bell-only' alarm. This kind of alarm usually consists of a control panel, a bell box (positioned on the outside of the house) and movement sensors. For added security, and at an extra cost, you can incorporate magnetic contacts for doors and windows, and a panic button. Such a system costs around £200 and can be installed for around £500.

Bell-only alarms work in a fairly basic way. In short, you set the alarm (for all or part of the house) from the control panel. If the movement sensors are activated, or the magnetic contacts are broken, the bell sounds. The only way to silence the alarm is to enter your personalized code. Depending on your specific circumstances, you can also install infrared sensors for specific room shapes, for draughts and changes in temperature, and for garages and conservatories. If the alarm is activated, you will have to rely on neighbours or passers-by to notify the police if you are not at home.

MONITORED SYSTEMS

If you live in a remote area or have many valuable possessions, you might want to consider installing a hard-wired system, linked by telephone to an alarm-receiving centre. This kind of alarm is activated in the same way as a bell-only system, but alerts the alarm centre when activated, which then notifies the police. A professionally installed and monitored system can cost around £30 per month, often with a one-off registration fee. You may also be tied into a contract for up to three years. This kind of system does not overcome problems such as an engaged phone line or an intruder cutting the line.

FULLY AUTOMATED SYSTEMS

A third option is to opt for a fully automated home-security system. These are available at varying levels of complexity, and at the top end it is possible to use computer programming to make your home look lived in while you are away, to switch on lights in response to movement, to lock and unlock windows and doors at your command, and to receive a call on your mobile phone if your alarm is activated. You can even set up your video to switch itself on when an intruder is detected and to film him or her in the act. This area of home security is hugely complex, and it is essential to seek advice from a reputable installer and get several quotes before committing to anything.

CLOSED-CIRCUIT TELEVISION (CCTV)

This is becoming increasingly more accessible to home owners, with basic systems simply plugging into the scart input socket on a television. More complex systems use a separate monitor and can power more than one camera.

Audio and video door-entry kits are also worth considering, and start from as little as £180. These allow you to hear or see whoever is outside, and mean that you can decide whether or not to let them in.

Seek advice from your architect or a crime-prevention officer to make sure you install a system that suits your needs.

Lighting systems

Although an electrician will install the lighting system in your new home, it will be the architect or designer who helps you work out what lights to install and where to put them. Many have a particular interest and good knowledge of lighting, and will know how to achieve particular effects and what's new and appropriate for your style of new build or renovation.

Draw up a lighting plan at the initial design stage with your architect or designer. This should take into account both natural and artificial light, so that you make the most of natural light in the daytime. Where you position lights will be influenced by this, and by how your furniture is placed – bedside lamps, a desk light, working light in a kitchen, for example. All lighting should be flexible, designed to adapt to the way you use rooms in daylight as well as at night.

LIGHTS FOR OUTSIDE

Exterior lighting is increasingly varied. For example, you can buy bollard lights or recessed lights for driveways and parking areas, and you can significantly improve the lighting around the house at night-time as a security measure. For more on security see next section.

Create a warm, welcoming and secure approach to your home with exterior lighting. Install lights in the eaves of your porch, but avoid blinding glare and contrasting dark areas by going for low wattage (12 volt) bulbs. Make sure you can override automatic lighting with a manual on–off switch.

For paths and driveways choose from LED bricklights (durable glass brick-shaped lights, effective when set into low walls) or LED walkover lights in the ground. Alternatively, for a subtler and softer effect, solar lights planted randomly under or around shrubs give just enough light to define a pathway, less regimented than a line along the edge of the path.

Think about these at an early stage of planning and consider the impact well-thought-out lighting can make to both the approach and appearance of your house.

Effective lighting of the house exterior can significantly improve your security, and provides an opportunity to highlight elements of the building's design.

GUIDELINES FOR WORKING WITH ELECTRICITY

In January 2005 the Government introduced a new law that demands most electrical work in UK households is carried out by a 'competent' person (see below). If you want to do at least first fix yourself (see next page), you should have a good understanding of electrical installations, as well as technical knowledge and experience of wiring.

The only reliable guide to wiring and lighting is an up-to-date copy of the Institute of Electrical Engineers (IEE) *Wiring Regulations*, although these are difficult to follow if you have had no electrical training. Any electrician you employ should be registered with the National Inspection Council for Electrical Installation Contracting (NICEIC) or the Electrical Contractors Association (ECA), although of approximately 100,000 electrical contractors in the UK, just over 10 per cent are registered. This is largely because registration is expensive and has to be renewed annually. This situation may change as the Government tightens up the rules on who is qualified to carry out electrical installations.

ELECTRICAL SAFETY REQUIREMENTS

The following are included in Approved Document P of the Building Regulations, which deals with the health and safety of people in and around buildings.

- Anyone carrying out fixed electrical installations – that is, the wiring and appliances that are fixed in households in England and Wales – must comply with BS7671, the national safety standard for electrical installations. People carrying out electrical work in Scotland must be registered as competent to do so. However, there are no regulations governing electrical installation in Ireland, although the Register of Electrical Contractors of Ireland recommends that you always using a registered contractor
- To quote the Approved Document P (page 9, Electrical Safety; see www.odpm.gov.uk), the work must be 'suitably designed, installed, inspected and tested so as to provide reasonable protection against them being the source of a fire or a cause of injury to persons'.
- You are legally responsible for ensuring that the electrical work carried out in your home meets the required safety requirements and has been carried out and certified by a competent person, that is, by an electrician registered with a Government-approved body such as the NICEIC or ECA.
- In addition to the safety aspect, in new builds you must now comply with new regulations concerning facilities for the disabled, which govern, for example, the heights of sockets (see page 101).

BATHROOM LIGHT FITTINGS

It is advisable to have enclosed light fittings in a bathroom, in case a build-up of steam should cause a light bulb to burst. Switches must be pull-cord operated from the ceiling, and any wall-mounted switches must be operated from outside the room. The IEE's *Wiring Regulations* divides a bathroom into four 'hazard zones'. Check manufacturer's instructions on light fittings to make sure that you comply with the regulations.

- **Zone 0:** In a bath or shower, light fittings must be low voltage (max 12v) and be rated IPX7, which is protection when immersed in water.
- **Zone 1:** Above the bath to a height of 2.25 m, light fittings must be splash-proof. A minimum rating of IPX4 is required here. Fittings must be protected by a 30mA (milliamps) residual current device (RCD) – a fast-acting safety trip switch
- **Zone 2:** An area stretching to 0.6 m outside the bath and above the bath if over 2.25 m. Light fittings must be splash-proof, with a minimum rating of IPX4.
- **Zone 3:** This must have a ceiling-mounted switch. Provided there are no jets of water, there is no IP rating required.

The IEE regulations do not make specific reference to wash basins but general advice is to treat them as Zone 2.

NB The IP (ingress protection) code, indicates first, to what degree the fitting is impermeable to dust and objects, and secondly, how waterproof. The scale runs from X or 0 to 6 for objects and X or 0 to 8 for water.

When should the electrical work be done?

Your electrician should lay cables and install galvanized-steel back boxes in the solid walls as soon as the basic construction of the house is finished. This is referred to as the electrical first fix. It takes place before any plumbing and plasterwork or dry lining (plasterboard). The plasterer will then cover the cables and leave wall surfaces flush with the back boxes. (For any stud walls, or for fittings in plasterboard, plastic dry lining boxes should be installed after the plastering stage.)

The second-fix, which involves fitting the switches and sockets and connecting cables to the electricity supply, happens once all joinery is complete and the fixing of guttering and downpipes for rainwater is done. For an average three-bedroom new build, first fix will take about a week, although you should allow longer for more extensive properties or those with special lighting requirements. Timber-frame new builds are much faster to wire up than brick or blockwork constructions, since there are plenty of cavities and no masonry to drill through.

SUPPLIES

As far as timing is concerned, you don't want to take delivery of second-fix electrical supplies until after the plumbing and joinery are complete. Nevertheless, it is useful to buy a few samples of light fittings – for example, downlighters – before the second fix, just so that you can get an idea of how many you may need and how they will look.

LIGHT SWITCHES

Wall-mounted light switches come with anything up to six switches (gangs), or dimmers, enabling you to control separate lights from one place. You can also have two- and multi-way switches so that you can turn lights on and off from different locations – from the top or bottom of stairs, for example. White plastic is the most common finish for switches, and MK is often the electrician's choice. If you prefer metal – brushed steel, chrome or brass switches – be wary of cheaper models that might be thin and bend or snap when it comes to fitting them.

PLANNING LIGHTS FOR EACH ROOM

With the fashion for widespread use of downlighters, electricians favour at least two lighting circuits per floor, allowing not more than 1,000 watts per circuit. This is sensible for safety, maintenance and in minimizing loss if a fuse blows. You can easily reach a circuit's capacity with 20 downlighters in a kitchen and bathroom.

- **Downlighters** Current fashion is for many recessed halogen downlighters in the ceiling. There are two types, both of which look identical. Low-voltage versions each have a transformer built into the fitting and are more energy efficient than the mains-voltage varieties. They also offer a greater choice of fitting. A sealed variety is available and is especially good for kitchens, bathrooms and showers.
- **Pendant lights** These are good in large spaces or over a dining table. Beware of using them in low or small living rooms or bedrooms where the visual effect is to lower the ceiling. Keep them away from cooking areas in a kitchen.
- **Floor lights** Low lighting around walls can be warm and flattering. Creating a sense of space, they are especially good in living rooms. Uplighting reflects light off walls or ceilings into the room and makes a room look bigger.
- **Wall lights** These are generally one of three types. Floods cast light back on to a wall or ceiling; glowing lights emit a diffused light directly into a space; and spotlights project small, bright pools of light with sharp shadows. All are good in hallways where they are off the floor and out of the way.
- **Spotlights** These are versatile and easily mounted on tracks. Good for bedside lamps and highlighting objects, they offer an alternative to traditional picture lights. Avoid using them in the kitchen where they get greasy.
- **Task lighting** Desk, floor or table lamps should provide concentrated pools of light – typically five times the intensity required for the rest of the room.

Home systems checklist

HEATING AND VENTILATING YOUR HOME

☐ Do your research: there are a number of boilers and systems to suit different needs and it is one of the most important decisions you will make, so do not delegate.

☐ Choose an active ventilation system to keep the air in your new home from growing stale.

MAKING LIFE EASIER

☐ Consider installing a central vacuuming system.

☐ Plan control systems for all your electrical appliances, computer networks and home entertainment.

BUILDING REGULATIONS

☐ Plan your interior and exterior lighting, plumbing and soundproofing with consideration for Building Regulations.

☐ Make sure the plumbing and electrical engineers you hire are registered with the appropriate bodies.

EFFECTIVE SECURITY

☐ Plan your defence against burglars by combining the latest in hi-tech computer-controlled monitoring and alarm equipment with clever landscaping tricks.

8 Environmentally friendly self-build

An increasing number of companies are making considerable efforts to minimize the impact their processes have on the environment, and this is particularly true within the building trade. Interestingly, self-builders are often at the forefront of this trend, constantly demanding more environmentally sensitive options.

Don't be put off by the stereotypical green house, with wooden cladding, grass roof, thick walls and solar panelling. Happily, some of the most effective energy-saving measures (through orientation, enhanced insulation and draught proofing) also happen to be the simplest, cheapest and least conspicuous. Other major aspects of 'green' building are the use of environmentally friendly design and materials. There is some overlap between the two – they can complement and contradict each other – and it is particularly helpful to seek good advice.

House building in the UK is dominated by two technologies, brick and block and timber-frame (see Chapter 5). It is quite possible to build green with either technology and it is by no means clear cut whether one is greener than the other.

One of the benefits of building a house from scratch these days, is that you can address environmental issues on many different levels.

Brick and block

There are three areas to consider when building with brick and block insulation, airtight building and mortar.

INSULATION FOR BRICK AND BLOCK

The first and most important step in building an environmentally friendly house is to enhance insulation beyond the requirements of Building Regulations. This way you minimize the heat loss from your building, which should also mean that you use the minimum energy to heat it. One way of achieving this is to increase the amount of conventional insulation you use, such as glass fibre, although by doing so you would have to expand the walls to allow for the extra bulk. A better option is to use the most effective insulating materials you can find from the outset.

For example, it is possible to buy a variety of extremely strong, but light, concrete building blocks, known as Aircrete. Even the very lightest of these are adequate for the average two-storey house, but are so full of air bubbles that they provide ten times the insulation given by the heavier aggregate blocks. They do need careful design – incorporating some reinforcement and movement joints – but builders find them easy to use.

Additional alternatives include using the most efficient insulation boards, such as those incorporating foam. Although these tend to be more expensive than traditional insulation – mineral fibre, for example – they will pay for themselves by lowering your fuel bills considerably.

Brick tends to do very little for the thermal performance of a wall. A thinner choice of cladding (hanging tiles or timber) will allow for more insulation on the inside of the house. These can be used with 'solid' blockwork – there is no cavity, just extra-thick, strong but light blocks with some insulation added on the inside and some form of cladding on the outside.

You can lose heat through 'cold bridges' – that is, where a heat-conducting material passes through a wall and can cause condensation to form. Examples of this are: wall ties, lintels, window and door reveals, and floor edges. There are many products on the market that avoid these problems, such as insulated lintels and special liners that break the joint between the bricks and blocks around window and door openings. Insulation can be wrapped around or over cold bridges at these points and at floor edges.

AIRTIGHT BUILDING IN BRICK AND BLOCK

Brickwork and blockwork shrink as a new house 'dries out' and little cracks open up. The importance of airtight building has only recently been recognized in this country, and the average builder will need guidance and monitoring to achieve it. Make sure your designer and/or builder is briefed to consult the relevant best practice guidance referred to in the Building Regulations. The main joints to watch are: around windows, doors and loft hatches; where pipes and cables pass through walls or ceilings; and where timber and masonry meet (such as built-in floor joists and where the roof sits on the walls).

This last area is particularly important because the timber shrinks too; gaps not present at the time of construction will open up later, and it will not be possible to access them to seal them. Around windows and doors, use a sealant joint, and use sealant around pipes and cables. Use hangers or fixings rather than build timber into walls: the cracks that appear between walls and ceilings or partitions can then be resealed once the house has finished shrinking. Otherwise, coving will both hide and seal these gaps. Wet plastering the walls will seal up the cracks in blockwork, while dry lining will merely hide them unless the lining itself is sealed at all its edges, which is difficult to do. The only way to prove a building is airtight is to pressure-test the house and trace leaks with smoke trails.

MORTAR ISSUES

Any bricks or blocks laid in a strong cement mortar – the majority in the UK today – can never be reused. Reclaimed bricks can be used but it is a shame to lay them in cement mortar for this reason. Weaker lime-rich (or pure lime) mortars are more environmentally friendly and are of quite adequate strength in most cases. They tend not to be used, however, because they set more slowly.

Timber-frame

There are two areas to consider when building with a timber-frame: insulation and airtight building.

INSULATION FOR A TIMBER-FRAME

It is often stated that a timber-frame is inherently better insulated than brick and block because the frame can be in-filled with insulation. In fact, there is no significant difference between the most basic timber-framed brick-faced wall and an Aircrete cavity wall of similar thickness. Nevertheless a timber-frame house offers opportunities to achieve much better performance through the following measures:
- adding more insulation between the frame and the cladding (like a 'fully filled' cavity wall)
- using a thicker frame to allow for greater insulation (140-mm deep frame as opposed to 90 mm) and thinner cladding (such as timber boarding, render or tile-hanging)
- using more efficient foam-board insulation in the frame (you could achieve the same insulation with mineral fibre but you would need to allow for more space for it)
- using board insulation attached to the inside or outside of the frame.

AIRTIGHT BUILDING WITH TIMBER-FRAME

An essential part of most timber-framed houses is the vapour check – a plastics-based layer that stops humidity getting into the structure of the house. This also doubles as an air barrier. Sealing all the gaps is essential for dampness control and will make the house airtight. Wiring and pipes tend to go through this layer, so great care is needed. There are special airtight electrical back boxes available, otherwise careful taping and use of sealant should suffice.

Another option is to make a gap with battens between the plasterboard and the vapour check for all the pipes and wires to run in.

Timber-frame is often presented as the environmentally friendly choice because it uses a renewable, natural resource. But it isn't that simple. Timber comes from a variety of sources – some near, some far, some well managed, some not – which can be difficult to establish. The best thing is to look for a sustainable timber source that has been certified by a reputable body, for example, the Forestry Stewardship Council (FSC) and Pan-European Forest Certification Council (PEFC). These bodies raise awareness in consumers and help encourage improvements in forestry practices.

Green insulation

In terms of minimizing the environmental impact it's more important to use a lot of insulation rather than any given type of insulation.

Cellulose fibre, hemp and sheep's wool products are the most green. However, as purists will point out, these can contain additives such as fungicides and fire retardants. Glass fibre and rock fibre are good green types. They can irritate the skin, however, so you should use protective clothing and masks when handling them. (Whether they have more serious effects than this on the health has been contested fiercely and, so far in this country, inconclusively.) Suppliers of insulation boards have improved the manufacturing process to minimize its impact on the environment.

Windows and doors

There are two main arguments about how green windows are: what the frames are made of and how insulating the glazing is.

While it is true that timber windows will need repainting, some factory-applied stains will last a good many years and these can be washed down and recoated without rubbing down. With aluminium-faced timber windows, the facings protect the wood and remove the need for painting.

Plastic frames don't need periodic repainting. However, some older plastic windows are showing signs of ageing and it is too early to tell how the newer generation of PVC-U will wear. The manufacturing of PVC-U involves materials and gases

that are toxic to the environment. Technically, double glazing is not mandatory for new builds; however, to meet the Building Regulation requirements for U-values (or energy efficiency) in most cases it is necessary. Gas-filling and 'low-e' coatings further improve the performance and insulation. Argon, or the more expensive but more effective krypton, may now be put into the cavity between the panes. Triple-glazed windows also are becoming more affordable as Scandinavian manufacturers are penetrating the UK market. Although the added insulation is good, some home owners don't like the loss of light that they bring.

Front doors can be significant losers of heat, particularly through the thin timber panels, and because they tend to lose their shape and fit poorly into the frame. Many high-performance window manufacturers make insulated doors. A porch or recess to keep the worst of the weather off the door will protect it, while an unheated lobby inside will help control air and heat leakage.

Solar power

'Passive' warming is when the sun warm a homes by shining through openings such as windows. 'Active' systems have been developed to intercept the sunshine falling on the walls and the roof and convert it into useful energy. Of these there are two main types:
- solar thermal, which uses sunlight to heat water
- solar photovoltaic, which converts sunlight into electricity.

Depending on the size of your house and the area covered by solar panelling, you can heat water for all your needs practically all year round.

HOW DO SOLAR THERMAL PANELS WORK?

Water passes through a solar thermal panel and is warmed by the sun. In most cases this warmed water transfers to a cylinder where it's kept until needed. The tank may need topping up occasionally – for example, during the winter months when there's less sunlight. Over the last few decades, solar thermal panels have progressed in design and are now robust, high-tech products. More than 40,000 households in the UK have solar water-heating systems.

HOW DO SOLAR PHOTOVOLTAIC (PV) PANELS WORK?

Sunlight creates an electrical charge, producing a current through a wire attached across the panel. The electricity is either used immediately or stored in a battery. Most PV panels are connected to the house's electrical distribution board via an 'inverter', which converts the direct current (DC) from the PV to alternating current (AC). When you use more electricity than your PVs can supply, a top up can be obtained from the national grid at the supplier's normal price; when the PVs supply more than you can use, the surplus goes into the grid and is bought off you by your electricity supplier at a different, usually lower price. Not all suppliers buy back and the price they pay varies. This arrangement is 'grid connected'. Some users, usually those who are remote from grid connections, connect their PV to batteries. This can be very efficient if the household appliances are DC. This is referred to as 'off grid'.

Recently, PVs have increased in efficiency and variety, while prices have fallen with the development of a mass market. PVs can be coloured, translucent, even flexible, and can be supplied as tiles, shingles and imitation slates as well as panels.

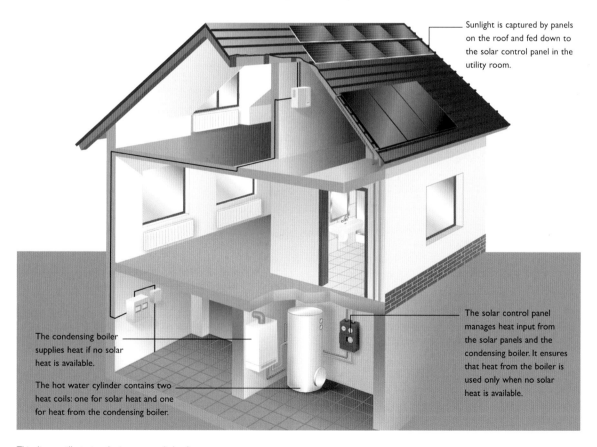

Sunlight is captured by panels on the roof and fed down to the solar control panel in the utility room.

The solar control panel manages heat input from the solar panels and the condensing boiler. It ensures that heat from the boiler is used only when no solar heat is available.

The condensing boiler supplies heat if no solar heat is available.

The hot water cylinder contains two heat coils: one for solar heat and one for heat from the condensing boiler.

This diagram illustrates the journey made by the electric current as it passes from solar panel to the solar control panel.

WILL THEY WORK ON MY HOUSE?

For either solar thermal or PV panels, a south-facing pitched roof is ideal. However, panels work over a surprisingly big range of angles; on a north-facing roof some PVs will achieve 60 per cent of their peak performance. Shadows passing over (such as a chimney or aerial) will reduce the performance of a thermal panel. Some PV panels stop charging completely if a shadow breaks the circuit but more tolerant panels are available. Obviously, both are more effective at sunny locations, although panels of both types can pick up a worthwhile amount of energy from an overcast sky.

Solar thermal panels are most cost-effective in households that use a lot of hot water. Generally, a hot-water cylinder is required; a few combination boilers accept preheated water from a solar panel but in most cases the panel needs a special, twin-coil cylinder. One coil collects heat from the solar panel, the other from the boiler. This is supplied as part of a normal installation. You can also fit a preheat cylinder to feed into your existing single-coil cylinder.

Models of either type of panel can be used as a replacement for the roof finish. This is referred to as 'building integrated' and both saves money on materials and attracts a bigger grant (see below). This is particularly useful with PVs as they tend to cover a larger area, and the manufacturers encourage this by producing them in the form of slates or tiles and by supplying preformed underlays that are fitted with the panels to form a weatherproof layer.

Remember that your build will need to comply with planning permission and Building Regulations.

DO THEY NEED PLANNING PERMISSION?

Whether or not solar panels need planning permission is debatable and there are examples of cases that have gone both ways. The decision is likely to be dependent on how far the panels will protrude from the roof and your local Planning Department.

DO THEY NEED TO BE PASSED BY BUILDING CONTROL?

Aspects of solar panels, such as their weathertight properties, structural loading and being part of a heating system, fall under Building Regulations, so your local Building Control Department should be informed.

WHAT WILL IT COST?

The initial costs of obtaining and installing a solar thermal panel are from around £2,500 to £3,500 and upwards depending on the model. These prices are for a professionally installed system – DIY systems are cheaper. Properly sized, specified and set up, they can pay for themselves in around 10–15 years, Recent Department for Trade and Industry (DTI) figures show that householders can save 40–60 per cent on their water heating bills, again depending on the fuel system it is replacing. PVs are expensive for the energy they produce. For a typical household the cost of installation could be around £12,000–14,000. The Government issues grants through its Clear Skies initiative to help with the cost – around 50 per cent of the cost.

Wind turbines

Windmills and water pumps have been part of the landscape for hundreds of years. Most of them have gone out of service in recent years – replaced by fossil-fuel-reliant machinery – but many electricity companies are now using wind farms to supply some of their customers' demands for electricity. On a smaller scale, there is also a range of turbines available that are compatible with the demands of a single house.

HOW DO WIND TURBINES WORK?

The wind turns the blades of a turbine attached to the top of a thin steel or concrete tower. The resulting spinning motion drives the gearbox and electric generator stored inside. The energy is used to power the building's electrical needs, with any surplus energy being stored in batteries for use when there is less wind. It is a good choice if you live in a remote and windy spot. However, for constant power, you will need another source of electricity as a back-up on calm days.

DO THEY SUIT ANY LOCATION?

There has to be enough wind, and it should be noted that obstructions severely slow the wind down, so built-up areas are rarely suitable sites. Even one single building up-wind of the turbine can cast a long shadow or create turbulence, both of which will reduce its efficiency and longevity. The ideal position is on a smooth hilltop. The wind moves faster higher up, so the taller the better.

WILL IT GENERATE ALL THE POWER I NEED?

The power from a wind turbine varies with the wind speed, which presents two problems. First, most households need a supply at a constant voltage. Secondly, you don't want a supply that shuts down for days, or even weeks at a time. Generally the turbine is connected to batteries, which are charged in order to supply the house with a constant voltage, and any excess energy is stored for use when the turbine shuts down. Batteries are expensive, however, and need a large capacity to soak up the power generated during long windy periods. Excess power can be stored in other ways, such as charging storage heaters, immersion heaters or heating hot water cylinders. Solar PV panels can complement wind turbines, as windless weather tends to be sunny. Even so, in many cases it makes sense to have a back-up supply, such as a diesel generator, to top up the batteries. This can be set up to start when the batteries drop below a certain level.

WILL I NEED PLANNING PERMISSION?

Yes. Although wind turbines fall in line with the Government's stance on encouraging renewable energy, your local Planning Department will need to know the exact location of the turbine and its size. The main issue usually concerns how visible the wind turbine would be and whether it would be a problem in a built-up area. Small wind turbines are fairly inconspicuous.

ARE WIND TURBINES NOISY?

According to the British Wind Energy Association (BWEA) a wind turbine is no more noisy than leaves on a tree being blown in the wind, and you cannot hear the swish of the blade from 10–20 m away.

WHAT WILL IT COST?

On some remote sites (which may also happen to be open and windy) it is cheaper, long term, to install a wind turbine than to get a grid connection to the mains. However, a grant is available through the Government-run Clear Skies initiative of up to £5,000. Turbines vary in price and quality and the likely replacement cost has to be set against the benefits of free electricity. A cheap turbine may be a false economy, especially on an exposed site where the unit gets a serious battering from the weather. Expect the supply and installation of a system complete with standby generator to cost between £15,000 and £20,000.

Earth-sheltered houses

There are about 60 earth-sheltered dwellings in the United Kingdom, and 6,000 North American earth shelters have been constructed in the last 25 years. Many more exist around the world, including Scandinavia, Russia, Australia and the Far East. These buildings are constructed underground or covered with a layer of earth, which provides excellent insulation, and which can be planted with vegetation to enhance conservation. They can be built in the countryside or in towns. The simplest method is to build into a hillside, so that three sides of the dwelling and the roof can be buried. To maximize the power of the sun and the insulating qualities of the earth, you could fit solar panels.

HOW DO THEY WORK?

In Britain, the underground temperature stays relatively constant at around 10°C, especially at 8 m deep. These homes tend to be very airtight, and many use mechanical ventilation (MV) to ventilate their inner rooms. A further logical step is to include heat recovery with the ventilation system, which reclaims heat from the outgoing air (see page 174). If the glazing is pointed south, it will catch the sun. While the temperature stays stable, the heat stays in

the structure and keeps it warm during colder weather. As a result of these measures, one example, the Berm House at Caer Llan, Wales, has no requirement for heating at all. The temperature in this house varies between 18°C in the winter to 22°C in summer. Other earth-sheltered homes have also achieved impressive energy savings.

DOESN'T LIVING UNDERGROUND LIMIT THE HOUSE DESIGN?

Architects of earth-sheltered homes have worked with a wide variety of concepts, from traditional to ultra modern, and from rectangular plans to the 'wheel' plan of Mole Manor in Westonbirt, Gloucestershire. Many have only one aspect, with three sides of the house facing into earth; but some peep out of the ground to the rear, or use a variety of means to allow light in from above. Openings facing upwards collect much more light than those facing sideways, so the problems of lighting the inner spaces are not insurmountable and present great opportunities for both dramatic and soothing effects. Light can also enter horizontally through internal glazing or with the use of a tube that redirects natural sunlight into a room.

ARE EARTH-SHELTERED HOMES ENVIRONMENTALLY FRIENDLY?

The principle of an earth-sheltered home is to work with nature, using light, space, vegetation and water to the best possible advantage. The insulation is greatly improved if large windows are at the front to maximize the sun's rays. Such builds are less environmentally friendly in terms of their construction. The walls have to be strong enough to support an earth-laden roof and so a large amount of concrete has to be used (more than for a traditionally built house). However, an efficient example will counter this through the energy that is saved once it is up and running. The Building Research Establishment (BRE) estimates that energy savings can be around 20 per cent more than a conventional dwelling.

IS IT EASY TO GET PLANNING PERMISSION?

Experiences of gaining permission have varied considerably. The Hockerton Housing Project –

Environmentally friendly homes can also be striking in design as well as practical.

recommended reading for anyone contemplating earth-sheltered building – gained permission because it was a green development. Others have battled for years for permission to build in beauty spots. Despite the house having a minimal effect on views, the presumption against new dwellings in the countryside still stands, and planners are as likely to reject a proposal in an old quarry as anywhere else for this reason. Compared with conventional designs, unusual houses tend to be considered more carefully by planners and not every council will be sympathetic. It's best to contact your local planning officer for advice before buying land.

WHAT ABOUT BUILDING REGULATIONS?

Building Regulations aim to protect the health and safety of the inhabitants of a building. There is little in the regulations specific to earth-sheltered homes and the principles are simple enough to apply. The regulations with particular relevance are as follows:

- **Structure** The shell is a major exercise in civil engineering and design work must involve the services of structural engineer.
- **Conservation of heat and power** Some non-standard calculations may be required to prove that your home will be energy efficient.
- **Fire safety** Regulations prohibit means of fire escape through another room in many cases, so inner rooms facing the earth may need a passageway to the front door.
- **Ventilation** There is a minimum amount of opening required when one room is naturally ventilated through another; mechanical ventilation is an acceptable alternative.
- **Resistance to moisture** The measures used to keep water out will be scrutinized closely. There are many efficient ways of waterproofing underground, which should be tougher than polythene sheeting. As earth-sheltered buildings have physically to hold back the ground water, the walls will have to be strong enough to take the extra pressure.

ARE EARTH-SHELTERED BUILDS EXPENSIVE?

Structural earth-retaining walls are expensive, as are the waterproofing and the high-specification structural glazing systems that tend to be used.

THERMAL MASS

A 'thermally massive' house is built with material such as concrete that absorbs excess heat – from cooking and sunlight, for example. This heat would otherwise be lost through opening windows or doors. Instead the heat is stored and re-emerges when the rooms cool, reducing the load on the heating system. However, a thermally massive house needs more heat to bring it up to temperature, such as on a cold winter's morning. If everyone is out all day, the house will cool down and waste all that heat you pumped into it. In hot weather, the house will tend to stay warmer at night.

A 'thermally lightweight' house, in which the walls don't absorb the heat, responds quickly to energy inputs, and so needs less heating energy to warm it up. However, there are disadvantages: it is prone to overheat when cooking or when the sun shines in; take away the heat inputs, and it cools relatively quickly, making built-in shade and rapid ventilation more critical requirements.

Thermal mass can increase your heating bills rather than reduce them and is only useful when:
• you have good access to, and are taking advantage of, solar gain
• you use a poorly controlled heating system (such as low-tech wood-burning stoves)
• you want your house to maintain steady temperatures.

Carpet, timber or other lightweight coverings over a thermally massive floor greatly reduces its effectiveness, and studies have shown little extra benefit in making massive materials more than 75 mm thick.

Some believe that extra-thick plasterboard provides quite enough mass, at least to soak up daily temperature swings. Others believe that the most efficient house is a super-heavyweight one, an earth-sheltered house being the ultimate.

However the energy saving should be much more efficient in an earth shelter than in a conventional house and very little external decoration is needed.

Straw bale building

Building using straw bales for walls is increasing in popularity among those keen to safeguard the environment, build their own homes, or simply save money. Straw bales create a robust and safe structure, which keeps the building warm in winter and cool in the summer. It has the added bonus of good soundproofing. There are over 100 straw-bale structures in the UK, ranging from outside toilets and sheds to a three-bedroom house and theatre.

WHAT IS STRAW-BALE BUILDING?

The art of construction mainly follows conventional methods of building, but standard bales of straw are used for walls instead of bricks or blocks. They are covered inside and out with traditional lime or clay render, which is lime-washed for additional weatherproofing. The bales also have to be protected from the wet and kept off the ground. It is quite typical to have the first layer starting at a

height of 225 mm. Traditional self-draining foundations of local stone, lime mortar and rubble are normally used, as are old car tyres filled with rubble. The latter increases the greenness of the build by recycling waste, and has the added advantage of providing a damp-proof course.

CAN IT HAVE AN INTERESTING DESIGN?

Compacted straw is very versatile and flexible, as is the traditional lime render. Walls can be built without corners so the design can stretch an architect's imagination. However, one stipulation is that roofs must overhang by at least 450 mm to protect the walls from rain. Straw bales are rarely used for internal walls because they take up a lot of space. Wattle and daub, or timber, are used instead, unless you want added soundproofing or insulation.

HOW SAFE IS A STRAW-BALE HOUSE?

It may seem that building with straw would be a major fire hazard, but tightly compacted and covered in a fire-resistant material such as plaster or clay it should be safe. Structurally, straw-bale buildings have proved to have far greater capabilities for structural loading than is required. The main enemy of a straw-bale building is water: if it penetrates the walls they will rot. This must be taken into consideration during construction. Overhanging roofs, well-draining foundations and effective damp proofing should be incorporated into the plans.

HOW ENVIRONMENTALLY FRIENDLY IS IT?

Straw is a renewable material and the bales provide super-insulated walls. Straw-bale builders use as many natural and sustainable materials as possible (see above) and adopt old traditional building methods to lessen the impact on the environment. If a structure is no longer needed the straw can be composted down when dismantled.

IS PLANNING PERMISSION HARD TO GET?

As with a conventional house you will have to get planning permission. Whether this type of building is accepted or not depends on the local authority. The building will also have to pass Building Regulations.

WHAT IS THE COST?

A three-bedroom, two-storey straw bale house can often cost about the same as a conventional modern house. It depends on the other building materials used. But the straw walls themselves will be cheaper. As an example, the brick-and-block materials for a conventional house will cost around £10,000, the 400 straw bales needed would cost around £600. You can reduce the cost by using recycled materials and, as the construction is a less technical process and is in fact fun to do, friends, family and other volunteers can help with labour. Therefore you can save on materials and labour costs. However, long-term savings on fuel are of the greatest benefit financially.

Straw bale buildings are cool in the summer and warm in the winter, which means that you can make great savings on fuel to heat your home.

9 Where to live

Once you've decided to self-build you'll need to consider where you're going to live while the construction is taking place. For most self-builders, their equity is tied up in their property so a move is a must.

Temporary accommodation

If your project has been scheduled to take six months or less you might be lucky enough to have family or friends who are kind enough to put you up. Make sure they are aware that building projects can go awry at any time: what might have been six months when you first moved in could stretch to nine or even 12 months if the project gets drastically delayed. There are a number of other options to consider.

RENTED PROPERTY

Most self-builders opt to move into rented accommodation for the duration of the build. You'll need to work out the cheapest and most practical way of doing this. For example, if you currently live 50 miles away from the building site and you want to be involved on a daily basis, then renting a property on the doorstep will be practical and can save on petrol costs. However, if it is expensive to rent in the area you'll need to work out whether living further away from your building site is more cost-effective. Furthermore, unfurnished property is generally cheaper to rent than furnished by around £50 a month, although it is worth investigating. It is best to calculate how much you would pay to store your furniture and whether you still make a saving by doing so.

SHORT-TERM RENT

If your house can be built in less than six months you'll need a short-term solution. It's unusual for landlords to let you sign up for less than six months unless you opt for a serviced property designed for short-term lets. This is generally more expensive than a typical six-month let and is charged on a weekly basis. However, the standard of property is high, with bills, cleaning and gardening services included in the price. It is worth remembering that anything could happen to increase the build time so it might be safer to guarantee your temporary home for a six-month timeframe to avoid having to move again. If tight on budget, and not expecting to rent for long, you might consider getting any children to share bedrooms. Obviously, a two-bedroom property will be much cheaper to rent than a three or four-bedroom house.

CARAVANS

This accommodation option may not suit everyone's lifestyle, but there are some considerable advantages to living in a caravan. Bear in mind that you can only live on the site of your build if you are going to be directly involved in the building project, and remember to consider where the caravan will sit in relation to the building footprint. You will not require planning permission for a caravan provided you are fully engaged in the project in some way, either physically or through project management.

Modern caravans have fully insulated walls, floor and roof, and are heated via a warm air system powered by electricity or gas. You'll have all the basic home comforts but on a smaller scale: storage space, a washroom, shower facilities, fully equipped kitchen, a fridge/freezer and a microwave in some cases. A four-berth caravan is big enough for a couple with two young children but is more suited to holiday-making than long-term renting. For example, a four-berth caravan measures 5 m in length while a six-berth caravan is 6 m; some models have specific advantages such as a big bathroom or separate bed areas. Go for the largest caravan you can afford. Planning departments usually allow mobile homes of up to 18 m x 6 m (and 3 m high) before permission is required. But check with your local planning authority before you spend any money.

Costs

Rental costs are seasonal, even for long-term contracts. Generally, the longer the hire period, the greater the discount. Don't forget to allocate costs for storing your furniture. Hire charges will vary around the country and availability will also be affected by the time of year. You'll be able to negotiate bigger discounts in winter when business is slow, while in the summer you'll have a limited choice and will need to plan ahead.

Have a look at how hire costs compare with buying a caravan. Prices vary depending on age, although caravans do not depreciate as quickly as cars. A new four-berth caravan would cost in the region of £12,000, while a good second-hand one might cost around £8,000. Used caravans can be bought for £2,500, although older models with minimal comforts can sometimes be bought for as little as £200 on the Internet. You'll be able to buy caravan insurance through the hire company but this will probably only cover you for theft. Make sure you seek the appropriate cover for the contents.

Connecting to the services

Your new house will need to be connected to the services and you can run temporary services to your caravan. For example, a temporary pipe might run from the mains water and you can buy a transformer for electricity. If your waste water cannot be directed to a suitable sewer system you'll need a waste-water container which is emptied by a specialist company. If there are no mains services for you to connect to, you can opt for a fully self-contained caravan with a gas bottle to run the fridge and oven, 12v batteries for appliances, lights, water pump and heating, and tanks for storing fresh and waste water until they can be emptied.

Organizing your possessions

Once you have decided where you're going to live make a list of what you are going to do with your belongings. List them in five categories: storage for new house; temporary accommodation; sell; charity

shop; and rubbish. This is a great time to rid yourself of anything that you have not used for years.

- New home: Will your furniture look right in your brand new home? Are your white goods the right size for your new kitchen? Have a look back at the plans for your new house and make a mental note of the spaces and the furniture you have to fill them. Get rid of anything that looks inappropriate. Look at your list of furniture and work out how much this might cost to store. You may find that the value of some items is much less than the cost of storage and it is cheaper to lose them. For example, the average lounge suite takes up 3.7 m² of space and will cost around £20 per week to store. Over six months, this will cost £520.
- Temporary accommodation: Check to see which white goods are already in your rented accommodation. Sometimes you'll be given the bare minimum (cooker, fridge) or if you are lucky there might be a freezer, washing machine and even a dishwasher.
- To sell: If you've got the time and you've got some items that you know could be sold for a good price, put an advert in the local paper, take it to a car-boot sale or put it up for auction on a trading website.
- Charity shops or rubbish: Don't throw away anything that might be of use to someone else. Some charities will collect furniture for free. Old furniture that cannot be sold or given away should be taken down to the local rubbish tip for recycling. There are receptacles at the municipal dump for glass, paper and old electrical items. These are all items that your council refuse collector will not take away.

Storage

Even if you move into unfurnished housing, you are still likely to have some items that need storing. If there is too much for family and friends to take on, you'll need to think about renting a storage unit.

Storage centres are purpose-built warehouses that offer a dry, secure and private way to store your belongings on a temporary (or long-term) basis. The units come in a variety of sizes ranging from around 1 m² to 100 m². Typically, a warehouse is monitored 24 hours a day by CCTV and protected from intruders by high security fencing, alarms and a fire- and smoke-detection system. It is your responsibility to load up your storage unit (trolleys are provided) and you'll be given your own padlock and key (which you usually have to buy). Access is allowed during the working hours of the centre, and in some case you can get in 24 hours a day. Costs can start at around £35 a week for storing the contents of a three-bedroom house, with white goods costing about £18 a week.

CALCULATING THE SIZE

It is difficult to work out how much space you'll need but as a very rough average a three-bedroom house takes up 9–13 m², while a four-bedroom house takes up 13–23 m². Some companies give online help through their size estimators. You select the furniture/appliance type and quantity from a list and it calculates the space you'll need. If you make rough calculations based on an average-sized house, don't forget to include your belongings in the loft and in the garage.

PACKING AND PROTECTION

A reputable storage centre will not have problems with damp and woodworm because the rooms are designed to be dry with an ample flow of fresh air. However, you'd be advised to protect antiques and delicates by wrapping them in bubble wrap (especially chair legs) and put a protective cover over them to avoid knocks and scratches. Leave items that you might need access to quickly near the front of the container and cover everything with sheeting. Storage insurance usually covers you for events such as loss or damage to goods caused by fire, floods and theft. Prices start at around £3 per £1,000 worth of cover.

Moving

Give your removal firm as much notice as you can and make sure you have booked their services at least two weeks ahead of time. The British Association of Removers (BAR) has a quality standard for furniture-removal companies, by which certified companies are required to define standards and procedures clearly to benefit the customer. Costs will vary depending on the size of your house and how far you are going. Once you've got a quote, check that everything has been included, find out how long it will take, whether it is guaranteed and what the insurance covers.

INSURANCE

Most removal firms will want to pack the contents of your home themselves and the insurance will cover you for loss or damage. Some companies will insure you if you pack yourself, but you will need to provide a full list of goods and their value. Things that cannot be insured include jewellery, software data, sets of furniture (if one chair of a set gets damaged, the mover only has to replace that piece, even if it doesn't match).

Using a professional company to move your belongings can save on time and anxiety, particularly if you are still overseeing the final stages of the project.

CHANGE OF ADDRESS
Check off the following list of people to notify:
• bank
• credit cards
• TV licence
• Inland Revenue
• insurance companies
• doctor, dentist, optician
• electoral register
• DVLA
• electricity
• gas
• water
• telephone
• mobile phone
• council

MOVING CHECKLIST

2–4 weeks before the move

- Start to notify companies of your change of address.
- Make arrangements for transporting pets.
- Start using up frozen and fresh food items so there is less to spoil.
- Have the car serviced.
- Sort out all your belongings, remembering the loft and garage.

1–2 weeks before the move

- Arrange for help on moving day.
- Make arrangements for children on moving day.
- Pack important documents and belongings in a 'valuables' box.
- Take books back to the library.
- Dismantle furniture that cannot be transported in one piece.

A few days before the move

- Defrost the freezer.
- Do last-minute laundry.
- Prepare an 'emergency' box with items such as change of clothes, toiletries, toilet roll, light bulbs, torch, screwdriver.
- Put valuables box in a safe place.

2 days before the move

- Clear out the fridge.
- Pack larder foodstuffs into a box and make sure boxes and bottles are tightly sealed.

One day before the move

- Disconnect electrical goods.
- Take down curtains if you are taking them with you.
- Pack a 'do not move' box with food for the moving day; kettle, tea, coffee, powdered milk, biscuits, cheese, fruit, bread rolls, plastic plates and cups.

On the day

- Strip the beds and keep the linen with you for when you arrive.
- Make a note of the meter readings for gas, water and electricity.
- Go around the house and make sure everything has gone and double-check the cupboards.
- Take your valuables box.
- Drop off all the remaining keys for the new tenants.
- Check that everything has been delivered and is in the right place.

10 Self-build case studies

Now you have read all about how to build a house, here are some real-life examples of self-builders' homes. On the following pages *Build It* readers share their experiences of building their home from scratch. As well as the joys and frustrations, they also share some tips to help you on your own journey to being a self-builder. Whether you are interested in timber-frame or brick-and-block construction, there should be something for you.

 POST-AND-BEAM TIMBER-FRAME (pages 201–203)

 A LITERAL SELF-BUILD (pages 204–206)

 DEMOLITION AND REBUILD (pages 207–210)

 BARN CONVERSION (pages 211–215)

 A CONSERVATION-AREA REFURBISHMENT (pages 216–219)

 BRICK-AND-BLOCK CONSTRUCTION (pages 220–223)

Post-and-beam timber-frame

Building a house of their own was the realization of a lifetime's dream for Bob and Ali Atkins. Having got on the property ladder at an early age, the pair were in an enviable position to self-build by their early forties. They only had a small mortgage to pay on their existing property and had been saving for the day when they would eventually build their own home.

FINDING THE LAND

By 1998, Bob and Ali felt the time had come to take the self-build plunge. Like many others before them, they started their adventure by looking for land. At the time they were living in a 300-year-old cottage in the tiny Lincolnshire village of Graby. Ali explains: 'We loved our cottage – especially its rural location – and decided we weren't prepared to buy a piece of land that we didn't like as much as our cottage site.'

They set about grading each piece of land they viewed out of ten, comparing it to their cottage, which they decided scored eight. Unfortunately, they had no luck finding anything that met their criteria or improved on their existing position. 'We really wanted at least a third of an acre in a remote rural position, with fantastic views,' says Ali.

After many months of searching for the perfect plot, a small miracle happened. The farmer, whose fields surrounded Bob and Ali's cottage, put a piece of his land on the market. The site, which had outline planning permission and was neighbouring their cottage, was exactly a third of an acre and, at

The clapboarding on the completed house is reminiscent of the New England-style properties that inspired its original design.

SELF-BUILD PROFILE
Names: Bob and Ali Atkins and children
Location: Near Sleaford, Lincolnshire
House size: 232 m²
Construction method: Douglas fir post-and-beam timber frame
Land cost: £50,000 (purchased in 1998)
Build cost: £230,000
Total cost: £280,000
Build cost per m² £948
Work started: April 2001
Construction time: 19 months
Current value: £370,000

£50,000, was within budget. Bob laughs: 'We just couldn't believe our luck and instantly snapped up the land. It seemed too good to miss.' Before the sale of the plot was finalized, Lincolnshire County Council insisted on an archaeological dig to ensure there was nothing of historical importance on the site. The cost of £1,200 was split between Bob and Ali and the farmer.

STYLE AND DESIGN

During family holidays to America, Bob and Ali fell in love with New England-style properties. 'I adored the post-and-beam houses that are so common out there – especially the vaulted ceilings, decks and clapboarding,' says Bob. With these elements in mind, they employed a small timber-frame company to draw up some initial designs for them. Bob knew that he wanted a vaulted lounge and hallway, and to have the kitchen in the centre of the ground floor. He was also adamant that the house should be constructed using the post-and-beam method. He comments: 'We feel that the kitchen is the hub of a house and where everyone seems to congregate. We wanted our new one to be in the centre of our home, with all the other rooms leading off it. Ali also needed a studio for her illustrating work.

It did not take long to formulate a rough design that Bob and Ali were happy with and, in December 1998, they submitted initial plans to South Kesteven Borough Council. Although satisfied with the drawings in general, the Planning Department made several requests that had to be carried out before

Bob and Ali opted for a vaulted lounge with light flooding in from windows on both sides of the room.

they would grant full detailed permission. First, the planners insisted that none of the structural timbers of the building – the posts and beams – be visible from the exterior. They wanted the building to look as conventional as possible from the outside. Secondly, they required the height of the new building to be no higher than that of the neighbouring house (Bob and Ali's cottage). This meant that a lot of soil would need to be excavated and the house positioned lower than the existing ground level. Finally, the new building's frontage was to be no larger than that of the cottage. This meant that the majority of the new property's accommodation would have to be at the rear of the building.

The archaeological dig was carried out in March 1999 and nothing was discovered. The amendments to the plans were made and, in July 2000, full detailed planning permission was granted.

INITIAL PROBLEMS

Quite early on, Bob and Ali grew uneasy about their package company. Initially, they had agreed on a price of £62,000 for the build, of which the couple paid £1,600 up front. Now, the company was asking for £72,000. Bob and Ali requested references and examples of other work the company had carried out, but were met with silence. Eventually, they decided to terminate the contract but the company refused to release their architectural plans for less than £6,000. Frustrated, Bob and Ali visited a local architect with their drawings, who agreed to draw up plans for £600, at which point the package company capitulated, releasing the plans for £600.

Bob and Ali took their drawings to another timber-frame company who quoted them £78,658 to build the frame. 'With hindsight, £62,000 was an unrealistic figure and should have caused alarm bells to ring – the second fee was much more achievable,' Bob says.

THE BUILD

The ground was broken in April 2001, and the foundations laid without a hitch. In addition to strip foundations, each load-bearing post of the timber frame also needed a 900 x 900 mm pile of concrete. A team of six men erected the timber frame in three days. Ali comments: 'This was a very exciting point in the build. Literally overnight we could see the size, shape and structural workings of our new home.' Once the frame was erected, it took a week to install the structural insulated panels (SIPs) that formed the structure of the walls. At this point the role of the timber frame was complete and the project became the responsibility of Bob and Ali's builder.

Having already spent a considerable amount of their accumulated funds, Bob and Ali put their cottage on the market to ease cash flow. Much to their surprise, they had instant interest in the property and the first people to view it put in an offer. Ali says: 'We accepted the offer, but were pretty daunted as it meant that we would have nowhere to live.' They decided to rent a small house on a nearby estate for the remainder of the build. Luckily, they were able to store their furniture, free of charge, in an empty barn belonging to the farmer who had sold them the land.

A GRINDING HALT

Bob and Ali had wanted to do a 'turn-key' self-build, which meant they wouldn't get involved with managing the project at all themselves. Unfortunately, they had appointed an unreliable builder and shortly after the frame and panels had been erected, the project came to a grinding halt. 'I think the builder bit off more than he could chew with our project,' says Bob. They encountered delay after delay: the brickwork was set back by four weeks and the roof by nine. In fact the second-fix didn't get underway until September 2002 – 17 months after the ground had first been broken.

Ali says: 'We got so frustrated. The builder became more uncooperative by the day and in the end it was as much as we could do to even get him on site. He appeared to have financial problems and completely lost interest in the house. Eventually, with no experience, Bob and Ali had to assist in managing the build. This became extremely stressful as they both had busy working lives. In desperation, they parted company with their builder, hired new subcontractors and Bob even did some of the work himself. He explains: 'I was getting so annoyed, that I just rolled up my sleeves and got on with it.'

Although the Atkins' build was very slow, it did give them the chance to plan what they wanted in detail – even down to the smallest fittings. 'On reflection, we recommend resisting the temptation to rush to finish a build,' laughs Ali. Once near completion, the building inspector came to view the property. 'We didn't think we had any worries regarding Building Regulations,' Bob says. Unfortunately, their builder hadn't been up to date with the latest legislation and the doorframes he had fitted were too small. According to current regulations, all new properties have to have doors that are wide enough to accommodate a wheelchair and these were not.

'I was furious. Not only had we already plastered everywhere, but all the doors and doorframes had been hand-made out of pine.' They had no choice but to remake and re-fit the doors and their frames – another setback they could have done without.

DEFINITELY WORTH IT

Bob and Ali moved into their new house in November 2002, a staggering 19 months after the build had started and four years from when they bought their piece of land. Bob laughs: 'Even though the build was very problematic, it was definitely worth it. We absolutely adore this house, and it's everything we ever dreamed of – I would even consider doing it again.'

Bob and Ali had set aside a £50,000 contingency fund on top of the £230,000 they budgeted for the build. They ended up spending the extra, but they don't regret it. 'If you're building the house of your dreams, and intend to live there for the rest of your life, a contingency fund is essential,' declares Bob.

BOB AND ALI'S TOP TIPS
- Don't be in a rush to finish your build – you'll appreciate the extra time when it comes to buying fixtures and fittings.
- Read the book *Timberframe* by Ted Benson.
- Make sure you get proper references before you employ a builder, and view the work he or she has done for previous clients.
- Work a contingency fund into your budget – you are likely to need it.

An open-plan living area radiating from a central kitchen was the focus of the family's design.

A literal self-build

When Andrew Gould's job presented an opportunity to relocate to Cornwall, he and his family could not refuse. However, when they came to look for a new home, Andrew and Karen, both in their forties, became increasingly frustrated. Although property prices were considerably cheaper in Bodmin than Essex, where they currently lived, they needed a home with a self-contained flat for Karen's mother.

Andrew says: 'It became clear that we weren't going to find what we were looking for and so self-build was suddenly a very appealing option.' Andrew believed that a successful self-build project would need a carefully planned budget. Before selling the family's three-bedroom, semi-detached house, he worked out exactly what they could afford. Working on the theory that the land and property prices in Cornwall were some 20 per cent lower than those in Essex, he calculated that he could build a 185 m^2 house for less than £110,000 – including land.

THE PLOT

It took Andrew and Karen just four months to find a plot that fulfilled their requirements. Their chosen site was part of a 10-acre development area, of which four acres had been split into 17 individual building plots. The Goulds bought one of these plots, with outline planning permission, for £35,000. The laying of roads and services was to be the responsibility of the developer, and Andrew and Karen considered that this, combined with having other self-build projects going on around them, would make their build run a little more smoothly.

This was not the case in practice, however. 'The developer was not interested in the self-builders' individual concerns: we had problems getting the services connected, which was out of our control, and there were delays from the outset owing to the slow clearance of the site and laying of the road,' Karen explains. 'In hindsight we probably shouldn't have bought a plot that was tied to a developer.'

PLANNING PROBLEMS

Since Andrew and Karen's plot had outline planning permission only, they had to gain detailed consent from North Cornwall District Council before the build could begin. Although the planners were not opposed to the detached house the Goulds were proposing, they did want the building to adhere as closely as possible to the housing in the surrounding villages. Andrew says: 'The planners imposed many restrictions affecting the external appearance of the house. For example, the front door had to be no more than five metres from the road, which was difficult on our corner plot. Also, we were not allowed to include any feature stonework apart from the front gable and we were not allowed to have gable ends or bow tops to windows. All of the walls had to be rendered with a brick plinth at the base.' The resulting house is much more featureless than Andrew and Karen had originally intended and it took five months before full detailed planning permission was granted. However, internally all their key criteria were achieved, including the self-contained one-bedroom flat.

THE PACKAGE

Having designed the footprint of their house themselves, Andrew and Karen took their plans to a timber-frame company that produced a package according to their plans. It included most materials such as a 90 mm timber frame, roof trusses, fascias, soffits, windows, external doors (including those for the garage), and staircases (including newel posts, spindles and hand rails – all cut and fitted on site). The company also supplied internal doors, flooring, architrave, skirting and all insulation materials. Andrew says: 'We were very impressed. I don't think we could have got a better package price than the one we had from them.'

SELF-BUILD PROFILE

Names: Andrew and Karen Gould
Location: Bodmin, Cornwall
House size: 325 m^2
Construction method: Timber frame
Land cost: £35,000 (purchased in 1999)
Build cost: £104,878
Total cost: £139,878
Build cost per m^2: £322.70
Work started: August 1999
Construction time: 24 months
Current value: £375,000 (estimated)

Despite various restrictions on design, the Goulds were able to keep the central stonework front gable.

THE BUILD

It soon became clear to the Goulds that careful budgeting was not the only consideration for a successful build. As the project progressed it was evident that a steadfast workforce was also essential. Karen says: 'Our first builder was recommended to us by the site manager of the developers next to us. He was well spoken, appeared to be very knowledgeable and inspired our confidence in him. Unfortunately we were mistaken, and things went wrong from day one.'

The Goulds were under pressure to release the timber frame from their suppliers because of the foot-and-mouth crisis, and time was not on their side. Much of the builder's work either took too long or had to be done again: he laid foundations that were short and his bricklayers were incompetent.

In the end, Andrew and Karen were forced to dispense with his services. Karen explains: 'The day we released this builder was the lowest day, but also the highest day – we realized we could start moving forward and make up for some lost time. We were really behind and had a massive bill for scaffolding which had been on site for 13 weeks instead of the nine we had budgeted for.'

WORKING ALONE

Karen employed a new builder to complete the blockwork and the rendering, and a plasterer to work on the first two floors. From this point forward they made the decision to complete the rest of the build themselves, only employing one or two contractors to assist with the completion of the project. The Goulds describe their project as a literal self-build. Andrew comments: 'We project managed, budgeted, purchased all materials, financially controlled and carried out most of the physical work ourselves.' With the exception of laying foundations, the timber-frame construction, slating the roof, exterior brickwork and rendering and some of the dry lining, Andrew and Karen completed the entire build themselves. This hands-on approach meant that Andrew and Karen were able to monitor the build's progress carefully.

They approached each room in turn and didn't leave it until it was ready for a floor covering. This approach helped keep them motivated during the more difficult days – a wise move considering there were 19 rooms, five built-in cupboards and a galleried hallway to finish. Andrew and Karen took a lot of care to eliminate possible sources of draughts during the build process and as a result the property

Andrew and Karen Gould built their kitchen to a high specification, including fitting all applicances.

To save money, the Goulds brought their carpets from their previous home into the new build.

was well within the recommended guidelines for thermal efficiency.

BUDGET REVIEW

The sale of their three-bedroom, semi-detached house in Essex almost entirely funded the Goulds' new home – there was a £21,000 overspend. Once work got underway, it did not take them long to realize that economies of scale would allow them to substantially increase the footprint of their house at very little extra cost. They reviewed their original budget and consciously overspent in order to create a larger, higher-specification home. For example, attic rooms were incorporated to provide a second sitting room, a study and a second gallery landing – all with beautiful views. Andrew and Karen also upgraded their bathroom sanitary ware and included a Jacuzzi bath in the master bathroom. They improved lighting and heating specifications and added powered garage doors.

FROM SMALL ACORNS

Andrew and Karen's build has been truly life changing. Not only do they have a beautiful new home, which would have been beyond their wildest dreams in Essex, but Andrew has now left his old job and started a construction company. Having lost confidence in the contractors who initially worked on his build, he decided to put his mechanical-engineering training to good use. He comments: 'The standard of finish that we achieved in our home made us realize that there is a very real demand for quality workmanship. We considered that we could put everything we learnt

during the process to good use and help others achieve the same standards in their homes.'

Andrew set up a partnership with a friend who had helped him with his second-fix work, and has a working collaboration with the timber-frame company that produced his package. The standard of finish he is achieving on other people's homes is now winning him more business. The Goulds admit that they would self-build again, but for financial gain and not personal use. However, they anticipate building a more modest property for their retirement, reaping the financial rewards and enjoying a good standard of living. Also, they want to be in a position to maintain one of their original objectives, which was to have enough collateral to relocate back to Essex if desired.

ANDREW AND KAREN'S TOP TIPS

- The most important part is to budget carefully – without a good budget you cannot monitor, reconcile and manage costs.
- Use a spreadsheet to give you the ability to monitor budgets against actual costs and to calculate your VAT return.
- A good budget gives you the opportunity to make informed decisions, good and bad, at every stage of the build.
- You don't know what you are capable of until you try. It was this realization that saved us most money during the build and achieved a standard of finish that has spurred us on to start a construction company of our own.

Demolition and rebuild

Ian and Susie Cornell loved their cottage in Little Baddow, one of the most desirable villages in Essex. They didn't want to move, until cracks appeared in the walls. The problem first became apparent in the detached garage block at the bottom of their drive in 1986, three years after they moved into the cottage. They were told that the cracks were the result of 'heave' – earth movement – and had the garages rebuilt. However, a year later large cracks started appearing in their living room and two of the bedrooms. The movement was monitored and it was decided that part of the cottage would have to be underpinned. The work was carried out and paid for by the Cornells' insurers.

SELF-BUILD PROFILE
Names: Ian and Susie Cornell
Location: Little Baddow, Essex
House size: 274 m²
Construction method: Timber frame
Land cost: £85,000 (purchased 1983)
Build cost: £245,000
Total cost: £330,000
Build cost per m²: £991
Cottage demolished: May 2000
Work started: September 2000
Construction time: 11 months
Current value: In the region of £1 million

Incredibly, the Cornells' elegant timber-frame house has a smaller footprint than their original cottage.

All was well with the house for the next 10 years, but Susie, who has multiple sclerosis, was finding it increasingly difficult to get around the house and surrounding site. They decided to sell up and buy somewhere that would be more convenient for them to live in. Just as they were about to put the cottage on the market, however, they noticed more cracks. Further monitoring took eight months, after which it was decided that a neighbour's tree was the likely cause of the problem. The tree was removed and additional repairs to the house completed in 1999.

A REPLACEMENT HOME

Much to the Cornells' infuriation, more cracks began to appear as the repair work was being finished. This time their insurers agreed to pay out the full amount for which the cottage was insured – £170,000. This made it possible for Ian and Susie to knock the house down and build a replacement home on the site.

By September 1999, Ian and Susie were finally able to start planning for their future. Working with a local architect, Brian Reeve, they measured the existing cottage and added on a percentage that they hoped might be acceptable to their local council's Planning Department. Ian says: 'One of the main criteria for us was to gain access to the rear of the property so that Susie could go straight from the car into the home. This meant losing the right wing of the cottage, which had been our lounge previously. We didn't want to alter the position of the house or lose any of the garden.

'We concluded that we would have to build a house with a vertical lift for Susie. Other requirements were an open-plan ground floor to make movement on Susie's scooter easy and a garden room to catch the sun. Upstairs we wanted a shower in our en-suite bedroom, with a sloping floor so that no shower tray was required. We decided that the two guest bedrooms could share a second en-suite bathroom, each with a separate locking door. We also decided that we would have the utility room upstairs. We thought, "Where's the logic in taking your clothes and bed linen downstairs to be washed?"'

UNDERFLOOR HEATING

'During this period we visited exhibitions and read self-build magazines, and became interested in incorporating underfloor heating,' recalls Ian. 'After

Ian and Susie Cornell's hi-tech kitchen has a central island as its focal point.

further investigation we decided to use this form of heating if we could find a local specialist, along with pressurized mains hot water and a condensing boiler. These would all be controlled from the utility room. The kitchen was another area we wanted input,' says Ian. 'We decided on a central island containing the hob and a food-preparation sink. As we steam a lot of vegetables, a built-in steam oven was also one of our priorities.'

3-D MODEL

Once Ian and Susie were happy with the layout they had a model constructed to give them a 3-D view of the house. They approached their local Planning Department in Chelmsford to seek planning permission. The new house would look very different from the old one and, initially, the council were unhappy with this, arguing that the replacement should be designed on a 'like-for-like' basis. However, the plot is well screened from its neighbours, and the other houses in the village were sufficiently varied in style. The Cornells and their architect were able to win the day. The new house would be a two-storey building (the cottage had been single storey), but the roof pitch would be just 20 degrees to minimize the increase in height. The footprint of the house would actually be smaller than that of the old cottage.

The double-height space and majestic arched window
give the dining area an almost cathedral-like atmosphere.

Planning permission for the new house was eventually granted in May 2000. The old cottage was demolished and the site cleared and levelled in September. Given the site's difficult history, it was clearly important to construct new foundations and floors that would be able to cope with any future soil movement. The solution was to drive piled foundations down into solid ground, with a ring beam over them and a beam-and-block floor on top of this. The groundworks were finished in December.

TIMBER-FRAME

Ian and Susie decided to have their new house built using a timber-frame inner leaf with an outer leaf of painted and rendered blockwork. The frame arrived on site in January 2001 and was erected by the start of February, ready for the roof tiles to be hung. The Cornells had initially intended to use Spanish slates on the roof but were advised that the low pitch would make this impossible.

THE FIRST-FIX

First-fix plumbing and electrical work was carried out during February and March. The concrete screed was also poured over the ground floor during March so that construction of the external block walls could begin. The blockwork walls were completed in April, after which the impressive floor-to-ceiling windows were installed. Susie says: 'The garden here is lovely and we have always looked on it as our "curtains". In the new house we wanted as much glass as possible to the front, so the windows and French doors had to be very attractive. We realized that PVC-U windows are maintenance free but we wanted wood on the inside. In the end, we ordered windows with a maintenance-free exterior and wood on the inside. We also managed to get the internal lift installed at this stage.'

During May the internal walls and ceilings were plasterboarded, a whole-house ventilation system was installed and the mains-pressure hot-water system plumbed in. The outside walls of the house were rendered in June, while inside, the stairs and central heating boiler were being fitted. During July, the kitchen was installed, the fireplace was fitted, the internal floors laid and doors hung. The external walls were also painted at this stage and the build was gradually transformed into a home.

August saw a final round of internal decoration, second-fix plumbing and joinery, and the construction of the external decking in front of the house. It was finished in seven months. Of the completed house, Ian says: 'We had never built a house before and have gained huge satisfaction and a vast amount of knowledge. We have also placed a time capsule, which contains many memories and pictures of the house and our lives, under the foundations. It's great to leave a mark on this beautiful landscape.'

The open-plan ground floor allows Susie to get around the house easily in her wheelchair.

Barn conversion

Ed and Marianne Seymour carried out an initial barn conversion, added a huge extension and followed this with a garden room. What's more, they plan to develop further. This is quite remarkable considering that significant barn extensions are prohibited.

It all began in 1999, when Ed and Marianne stumbled across their five-and-a-half acre plot by pure accident, during a visit to the Wiltshire village of Oaksey. Ed comments: 'It's amazing, you hear of so many people really struggling to find a site and we just spotted ours during a trip to the village pub.' Ed and Marianne bought the land complete with a complex of barns a year later. The site had planning permission for two barn conversions and the build of a new home. As an architect, Ed was keen to convert one of the barns into a functional home. In July 1991, after knocking down a large corrugated-iron barn and completing a stylish conversion, the Seymours moved into the smallest barn on the site.

THE INITIAL CONVERSION

This part of the transformation cost £80,000 – an amazing achievement considering they also renovated two long byres. They incorporated solid-oak flooring and introduced bespoke windows, underfloor heating and a wood-burning stove. Ed says: 'Using quality materials pays off in the long run. If you're thinking of the future, use materials that don't need a lot of maintenance.'

The small barn holds fond memories for Ed and Marianne. They lived in the tiny dwelling, which

This self-build project began with a barn conversion, which then had two subsequent extensions.

contained an open-plan living/kitchen area, an entrance hall, a bedroom and a bathroom, for five years. They left many of their belongings in London and grew used to a life with only the bare necessities. Marianne says: 'We thought about everything so carefully. The cupboards in the bathroom have glass doors designed to reflect the room and make it feel bigger. The bathroom suite and the bed in the room beyond were hidden behind doors so that the rooms looked like living space from the front door. After much thought, we even included caravan-style kitchen cupboards, with racks on the doors to cram in as much as possible.'

Despite having small dimensions, Marianne says the barn didn't feel too small. She adds: 'We kept the kitchen units at waist level so there was no visual divide and refrained from hanging pictures on the stone walls, to make the space feel open and airy.' The domestic situation got even more interesting with the birth of Ed and Marianne's daughter Ella. With no provision for a second bedroom, the hall had to make do and Ella's bed was hidden behind a curtain.

The Seymours employed clever devices to maximize on space, such as using curtains to divide areas of the house.

A CHANGE OF PLAN

Ed and Marianne had intended to live in the small barn for a temporary period and later move into one of the larger buildings. Unfortunately, the recession hit and in 1995 they had to change their plans. Ed comments: 'We borrowed some more money to build a house for resale on a patch of our land – we also sold one of our barns.' The barn was sold on the condition that Ed drew up the plans for its conversion and that the Seymours would have the final say on how the building looked. They were anxious that the conversion wouldn't contrast with the simple exterior of their home.

PLANNING PERMISSION

With their debts paid off and money in the bank, Ed and Marianne decided to extend their own barn. Unknown to them at the time, this was against the current planning regulations. Fortunately, the chief planning officer of North Wiltshire District Council discovered that the barn had been granted permission to extend 30 years previously. He was impressed with the quality and style of the original conversion, and figured that the initial permission could still stand. He was also convinced that the extension would be in keeping with its surroundings and an asset to the development. This series of elements enabled the permission to be granted – a small miracle.

Charles Uzzell, the council's planning officer, confirms: 'Although the chances of a barn being granted permission to extend are slim, it's worth checking with your local council. There may be exceptional circumstances that make the building exempt from current regulations.'

THE EXTENSION

After much deliberation, Ed and Marianne decided to extend the barn as simply as possible, choosing to make it about three times its original length. Ed didn't want to fall into the trap of trying to make the interior of the extension look old, but he was keen to establish some kind of continuity. He used the same oak flooring throughout the build and installed rafters at exactly the same width apart, stained to match as those of the original barn. Ed and Marianne then painted the plastered walls brilliant white to enable them to hang pictures.

From the outside, it is virtually impossible to tell that the property has been extended.

Although the extension looks very modern inside, it was actually constructed using simple techniques. Costing a total of £120,000, it consists of a breeze-block structure with traditional Wiltshire stone on the exterior. Luckily for Ed and Marianne they found a great deal of the stone on their site, which saved them a lot of money. The Seymours incorporated a kitchen, a large dining area and a suspended first floor to act as an office.

Edward and Marianne had initially incorporated a large overhang into the first floor, which leads into a dormer window, but decided it looked too conventional and ripped it out. Marianne comments: 'Architecture is such an expensive art but if something isn't right, and you've got the money, you have to change it.' They also included two extra bedrooms, an en-suite bathroom and a utility room.

By painting all the walls brilliant white and minimizing on clutter, the Seymours have created the illusion of space.

Ed knew that using quality finishes would make a tremendous impact on the success of a design.

The kitchen worktop is made out of zinc, and Marianne comments: 'The company was reluctant to provide the top for us because zinc marks badly, and they were concerned we would moan.' Ed and Marianne convinced the company that the opposite was true and that they liked the distressed look of the metal. They were also so impressed with the wood-burning stove they had installed in the initial conversion that they put another one in the extension.

The kitchen and dining area is floored with limestone. Ed explains this holds the key to their look: 'The floor was expensive, but sets everything else off beautifully. It makes all the fittings look exclusive, even though they're constructed out of painted MDF.' This proves that, with careful planning, a bespoke look can be created on a restricted budget. Ed and Marianne's style also hinges on restraint – there is no room for clutter and everything must have a home. There are invisible cupboards everywhere, keeping the architectural lines clean.

Internally, every element had to serve a purpose. For example, the wall around the kitchen acts as a very shallow cupboard that is perfect for tins.

THE GARDEN ROOM

A year after the extension was completed Ed and Marianne decided they wanted to develop further. They submitted plans to introduce a glass garden room and were surprised when the visiting planning officer informed them that they couldn't add an extension to a barn. Marianne comments: 'We explained that we had already extended once and the officer was astonished – he had assumed that the new area of our home was part of the original barn.' Ed and Marianne wanted to be able to open the doors of the garden room entirely, giving the

The original barn structure provided the best possible inspiration for the Seymours' subsequent development.

feeling that you were actually outside. Ed was concerned that sliding doors would be difficult to weather- and draught-proof but, after much hunting, discovered a German company that had the solution. Ed explains: 'They prefabricated the whole garden-room structure, incorporating a unique door-opening system and then shipped it over.' With a total cost of £30,000, the garden room was an excellent investment – Ed and Marianne even extended their heating, making it a living space for all seasons.

FUTURE PLANS

Five years have passed and Edward and Marianne have got the building bug again. They're in the middle of adding another extension to their home. It will include bedrooms and a second lounge with another wood-burning stove. They have also just bought a piece of land in Sri Lanka and plan to build an ultra-modern home on the site.

ED AND MARIANNE'S TOP TIPS

- Use a good architect – interview more than one.
- Make sure your architect is aware of what you like.
- Respect your surroundings.
- Internally, use one expensive material to boost everything else – we used limestone flooring.
- If you've got time, project manage yourself. Hiring tradesmen as needed will save a lot of money.
- Make sure your architect is aware of your financial limitations.
- Don't expect things to go to plan – people often get annoyed when builders don't produce on time and it isn't always their fault.
- Think about how you want your space to look and make provision to hide away ugly necessities.
- Use local suppliers – if you build up a good working relationship they are normally more loyal and reliable.

The choice of design means that it is virtually impossible to tell old from new from the exterior of the house.

A conservation-area refurbishment

It is not often that a house comes on the market in the popular picturesque coastal town of Southwold in Suffolk. It is even more rare for that property to be a reasonably priced Regency Grade II listed cottage.

SELF-BUILD PROFILE

Names: Nick and Yana Brooker

Location: Southwold, Suffolk

Style of property: Grade II listed fisherman's cottage

Period of property: 1690, with later additions

Property cost: £120,000 (purchased in 1998)

Conversion cost: £120,000

Total cost: £240,000

Work started: 1999

Construction time: Three years (with a planned break in the middle)

Current value: £470,000

Nick and Yana faced a number of restrictions in renovating their house, because it was Grade II listed.

It had been a long wait for Nick and Yana Brooker to find their dream house. Aware that most property in Southwold sold by word of mouth, rather than through an estate agent, Nick and Yana decided to rent for a while. They got to know the local community and this was how they came to find out about the fisherman's cottage. Back then, in 1998, it was valued at £120,000 but it was in desperate need of repair. All the ceilings were covered in Artex, with central light fittings, and many were bowed. There were cork floors throughout, and much of the wallpaper was at least 40 years old.

On entering the cottage from the street there was a small lobby, which opened on to a tiny living room. A staircase occupied most of the back wall of this room and led to the first floor. Under the stairs was an old boiler. Also on the ground floor was a dining room with a small kitchen leading off it. Upstairs were two tiny bedrooms off a lobby landing, a small shower room and several wooden doors. In one bedroom a wooden ladder led to an attic with a very low ceiling (about 1.2 m high). Nick said: 'We knew immediately that it needed a lot of work, but it was almost impossible to find a house with a modern interior in the town anyway.'

HUGE POTENTIAL

Yana explains that the poor state of the cottage was largely due to neglect, probably because it had been a holiday home for 25 years. 'What attracted us immediately was that we could visualize the potential to create exactly what we wanted: a modern and uncluttered home. Its position in the town, just thirty steps from the sea, was irresistible.' For Nick there was another attraction. He comments: 'On the way into town you pass *The Lord Nelson*, or as the locals call it "The Nellie", it's the best pub in town.'

When it came to renovating the property, Nick and Yana's priority was to create space and light, with just one recreational area downstairs, and a large bedroom upstairs. Nick explains: 'Nearly all the windows are on the front, and are south facing, so we needed to take advantage of that, and open up all the internal spaces. Both Yana and I are tidy people, and we liked the idea of a very minimal environment.

At the time there were just the two of us. However, we have now increased the bathroom area, with a small extension, and formed a second bedroom for our daughter Lollipop. We were greatly restricted by the planners as to what we could do because the property is Grade II listed.'

The renovation work took just over three years with a break in the middle. The work was completed by a local builder who had many useful and inspirational ideas. Nick and Yana had looked at several builders and the work they had done before choosing the right man for the job. The Brookers decided to split the work into two stages. They explained: 'We needed a break and couldn't live with it all in one batch.'

DAMP PROBLEMS

First of all came the necessary works. As with many old period properties, the fisherman's cottage had damp. One of the biggest problem areas was the wall facing the street. When the builders removed the plasterboard, they found thick ivy growing up the inside wall. They practically had to demolish the

The Brookers were keen to keep as many of the original features of the cottage as possible.

The open-plan design of the ground floor was an essential feature of Nick and Yana's refurbishment.

wall and rebuild it, with much new brickwork, in exactly the same style as the original and giving further support with steel reinforcing rods. This was no easy task, as much of the brickwork dated from 1690. Nick reflects: 'We discovered that the things you don't see are the ones that take the time and the money. I would estimate that we have spent well over £100,000 on the whole property to date. There were also practical difficulties for the builders owing to the lack of space.'

Included in their cost was not just the structural and maintenance work, but also the cost of modernizing the interior of the property. Nick and Yana wanted to protect the original features and made sure that all the products they used, such as dry lining on the walls, did not upset the existing structure. Any other modern finishes were placed on top of the existing ones so that future owners could remove them if they preferred a more period touch. This approach pleased both the planners and Nick and Yana's desire for sleek lines and a smooth finish.

The second stage of the build involved erecting a small extension on the back of the cottage and converting the attic. This allowed for extra space in the bathroom and the creation of a ground-floor shower room and toilet. Planning permission was relatively straightforward. This was largely because the planners liked the first stage of the development. However, they had one small problem over the size of the proposed structure. Nick explains: 'They would only allow the extension to be 1.3 m long. Of course this was a constraint. The planners decision-making was governed by the fact that we live in a conservation area, together with the close proximity of the buildings around us.'

The fact that the cottage is listed, and is in a conservation area, meant the Brookers also had to abide by the conservation bodies and follow traditional methods of construction. Nick says: 'For instance the preferred product for rendering this house is horsehair and lime, which has to ferment for six weeks. I understand that listed buildings must be

The two small bedrooms were converted into one to create space and let light into the cottage.

treated with respect and care for future generations, but sometimes that simply isn't practical.'

DIFFICULT TIMES

Nick and Yana lived in the cottage through all the building work, but it was far from easy. Yana comments: 'It was almost divorce. We were cleaning up every day as soon as we came in from work. It was a nightmare. No matter how tidy builders are, and ours were brilliant, rubbish and dust creep in.' On reflection both Nick and Yana admit they would have preferred to have converted before moving in. However, they had not carried out a building project before, and inexperience led them to believe it would be fine to live on site during the construction. Nick advises other renovators: 'You need to become a site foreman and project manager, even with little knowledge, just to help keep the job on track. You also have to realize that costs and timescale can easily grow. To carry out a project like this you need a lot of perseverance.'

But they are thrilled with the results. There is a seaside theme throughout the interior, much of it designed by local artist Serena Hall. Yana says: 'She has been an inspiration to us and we have bought several of her ceramics, which go so well with our decor.' Off the dining area is a compact galley kitchen: the units were given new fronts by the builders, and have brushed aluminium handles bought from a yacht chandler. The room opens on to a walled courtyard.

Yana and Nick decided to introduce a spiral staircase to increase the space. Yana explains: 'We got the idea from original shop interiors of The Arcade in Norwich. We also decided to make the fireplace a display area as we now have central heating.' The pair go on to explain: 'We are going to create a structure like a sailboat canopy in the garden. There will be a granite floor, large water feature and lots of lighting.'

Would they self-build again? 'Yes,' says Nick, 'next time with a Georgian property. However, we would still furnish it very simply.'

Space was tight on the ground floor, so Nick and Yana opted for a compact galley kitchen.

The Brookers applied the same coastal colour scheme throughout the house, creating a sense of continuity.

Brick-and-block construction

When Nigel and Bella Roberts saw an advert for a half-built house and land in Malmesbury, they decided to take the opportunity to build their own home. They had a good idea of what they wanted and worked with a design and package company to realize the dream. The head of the design team managed to take their ideas and turn them into the perfect house. Although they started with modest expectations, they ended up with a much larger house – almost twice the size of a conventional house. 'We would not change anything,' says Nigel.

SELF-BUILD PROFILE

Names: Nigel and Bella Roberts
Location: Malmesbury, Wiltshire
Construction method: Brick and block
Build cost: £154.972
Total cost: £339,471
Work started: 2001
Construction time: 21 months

Having gained planning permission and demolished the half-built house, building work began in September 2001. Nigel and Bella originally chose to build using reconstituted Cotswold stone bricks on the outer walls. However, they changed their minds once they saw a neighbour's house and opted for natural stone from Farmington. The quoins, the blocks of the outer corners of the wall, were also made from natural stone as were the door and window openings. The detail was an important feature of the whole house.

They opted for beam-and-block floors for better sound insulation and to minimize squeaky or bouncy floors. A long-span, concrete first floor reaching from one side of the house to the other, is supported by the exterior house walls. This allows for the upstairs floors to be supported on the first-floor slab. For the roof of the house, Nigel and Bella opted for distinctive, blue-black, textured slate tiles from Spain. To maximize their living space they decided to use the space under the roof, which gave them another storey to put bedrooms in.

Although they started the self-build process full of excitement and intending to have a stress-free

Nigel and Bella's house was a new build from scratch.
The end result was well worth the effort.

The Roberts' opted for beam-and-block on their build.

On the exterior wall they opted for natural stone.

build, early problems with their builder gave Nigel and Bella cause for concern. Nigel says: 'In hindsight we were a bit naïve. When our relationship with the builder turned sour our stress levels increased to almost unbearable levels.' Completion had been set for May 2002. By January 2002 it was obvious that things were not going according to plan. The roof tilers failed to appear on site on their due date and then did not start work for a further six weeks.

The project continued to suffer from delays and the Roberts' relationship with their builder weakened. The Roberts' postponed completion from May to September 2002, and then again to November. Bad weather in November delayed progress yet again, and the couple were unable to move into their new home until January 2003, having relieved their builder of his services twice! It took Nigel and Bella a further six months to complete the internal finishes themselves.

Despite the trials and tribulations, the Roberts are very happy with their new home. 'We've built a house that we love and that is perfectly suited to our family's lifestyle. It has been a massive achievement, and one that we are very proud of. Also, for me it has been the culmination of a lifetime of dreaming and planning,' says Nigel.

Financially, it has paid off handsomely, too. 'We went about £30,000 over budget and most of this cost can be accounted for by the external works. Our landscaping turned out to be significantly more ambitious than we had originally anticipated.

Beams are carefully placed in the right position for the blocks.

Making allowances for a dormer window is a complicated business.

As the slate tiling on the main house drew to an end,
work was mid-flow on the smaller building.

We planned our finances very carefully and had a contingency to cover any unforeseen delays.'

Would they do it again? 'That is a very good question,' says Nigel, 'and no, we do not plan to do this again. We always intended this to be our house for life, and we built it with that intention in mind. Having said that, if we ever saw a building plot that was just right, I imagine in a few years, we might do it again. But, leaving this house would be a wrench as we love it and have put so much of ourselves into it.'

Nigel and Bella maximized the space by building into the roof space.

The contrast of the blue-black slate against the golden limestone of the Farmington stone has created a stunning looking house.

11 Frequently asked questions

These questions originally appeared in *Build It* magazine during 2003 and 2004. Some details may have changed and we recommend that you seek your own professional advice.

Buying

HOW DO I VALUE A SITE?

Q My husband and I have made an offer on a plot that is in two parts. Plot one is about 565 m² and has outline planning permission for a four-bedroom detached house. Plot two is about 275 m², with no permission, and the plots cannot be purchased separately. Can you tell us the value of these plots and the approximate cost of building a four-bedroom detached house?

A It is not possible to value a plot of land purely based on its size. Plots for single houses can vary from £30,000 to £300,000 depending upon the location and desirability of the site. The cost of building a four-bedroom house depends on whether you are going to ask a builder to do all of the work or whether you are going to do some of the work yourselves. But, for budget purposes, work on £750 per square metre plus £10,000 for paths, drains, fencing and so on.

PRIVATE SALES

Q We've found a plot of land that the proprietors are selling privately. They are sending details to various interested parties and have said they will decide who to sell to when they have received all the offers. They advise that offers should be in excess of £125,000. How much more should we offer? We were thinking of £130,000. Also, do we need to get solicitors involved at this stage to carry out land searches or are we better off waiting until we know if we have been successful?

A It depends how formal the tender procedure will be – this sounds very informal. If you can afford to, offer slightly more than £130,000 – say £130,500. You might beat other potential purchasers who are thinking along the same lines as you. Qualify your offer by letting them know that you are really keen and hope to be allowed to resubmit if a higher offer than yours is made. Any offer you make should be 'subject to contract' and your solicitor should become involved at that point. However, it will do no harm to advise the vendor of your plans and mention in your offer that you have a solicitor lined up – this will show you're really committed.

HOW MUCH LAND DO I NEED?

Q How much land is needed to build an 185 m² house?

A It depends on how big a garden you want. Assuming your house is on two storeys, it would be reasonable to look for a plot about 545 m². That would give a front garden of 6 m long, with 18 m at the back.

RIGHT TO BUY

Q I am hoping to buy a plot of land subject to planning permission being granted. What can I do to make sure the vendor doesn't sell it to someone else in the meantime, or build themselves, once permission is granted?

A The usual way forward is to enter into a conditional contract with the vendor, where you exchange contracts but don't complete until planning permission has been granted. An alternative is an option agreement, under which you offer the landowner a token sum for the right to buy the site at any time in, say, the next two years. You agree to use your best endeavours and expense to obtain planning permission. Should you exercise your option to buy the site, you'll pay a certain sum – the amount agreed when signing the option – which is usually a percentage of open market value (something like 85 per cent). The discount reflects your time and costs of getting permission and any up-front fee you've paid for the option.

RANSOM STRIP

Q We are in a quandary over how much to offer for a potential building plot at the end of our neighbour's garden. A strip at the end of our garden would provide the only access, and we think this 'ransom' strip holds the key to any planning. We believe that we should offer 60 per cent of the full value. Our neighbour's estate agent says we should pay the full asking price. Can you comment?

A If the site cannot be developed without your land then, in theory, you each have two parts of a development site and it would be appropriate to split the development value 50/50. The development value is the plot value less the existing use (garden) value. Your 60 per cent of full value looks generous. No developer would pay the full price for a potential plot with a ransom strip. If it has no planning permission or there's no agreement with you over access, a developer wouldn't pay much more than existing use value.

CONDITIONAL CONTRACTS

Q We are searching for a plot of land locally and have found a house nearby with a large rear garden. If the owner agrees to sell the land, how should I progress the situation? Assuming there is no planning permission, if I offer to pay costs to obtain consent and we agree a price for the land, is there any way I can guarantee that the vendor honours our agreement and sells me the land and at the agreed price?

A You can enter into a contract to buy, which is conditional on you obtaining the planning permission you want. You agree a price and a time frame to allow you to get permission, and then exchange contracts, with completion delayed until permission is granted. No permission, no completion. It sounds simple enough, but solicitors like to tie these things up pretty tight and will want to determine what

constitutes an acceptable planning permission, what happens if it's refused (such as allowing time for an appeal) and what happens if there's some drastic shift in the market during the agreed time period. Your risks, therefore, are your legal and planning costs. There's also the question of a deposit to negotiate and whether this should be returnable. Sometimes the price is left flexible, depending on what kind of permission is obtained, or a figure is agreed with an uplift built into the contract. For example, you may end up getting permission for a five-bedroom house when initial inquiries to the planners led you to think all you'd get is a three-bedroom bungalow. Proceeding without a contract is very risky. You could spend time and money in getting permission only to have the owner put the plot on the market in the meantime.

CLAW-BACK CLAUSE

Q We've found a plot of land for sale and in the details there is a reference to a 'development claw back' clause. This requires payment of 51 per cent of any net development value in the event that planning permission is given for a purpose other than agriculture. What does this actually mean, and how does it affect a self-builder looking to erect a single dwelling on the plot?

A The clause means that any purchaser obtaining planning permission would have to pay an additional sum to the current vendors. The amount would be just over half the development value – that is, what the site is worth with permission minus what it was worth without it. In other words, it's just a way of the current owners ensuring that they benfit from any permission being given. As the site does not have planning permission it's debatable whether you should buy it at all: if it had had good prospects the owners would have applied for permission themselves and sold at full development value.

DRAIN ON RESERVES

Q I'm hoping to buy a plot of land that has several drains running through it. How do I find out whether they could interfere with my plans to develop the land?

A Drains can be surface or foul water, and can be public or private. Sometimes they can be diverted. Check first with the drainage authority – either the local council or water board – who should have records of the drains. If they are public drains, establish their size and whether there is a no-build exclusion zone around them. If the drains are private, check the deeds to see if anyone has a specific right to enter the site for maintenance. If so, you might be able to divert the drain if it's in the way, although the owner of the drain might want compensation for any change or disturbance. Your vendor should have resolved any issues prior to selling (unless you are buying at a price that reflects the uncertainties). If you can't get clear answers quickly, but are otherwise keen to buy, consider making an offer 'subject to satisfactory resolution of drainage issues'. Failing that, discuss the issue with a local chartered building surveyor, who could advise on the practicalities and costs of dealing with the situation.

Planning and budgeting

PROPERTY DEVELOPMENT

Q I work in an architects' practice and two colleagues and I are considering building houses and selling them when they are complete. Is this still classed as self build and would we qualify for a self-build mortgage? Or would we have to present the idea as a business case and set up a small business? Also, would we be eligible for value-added tax (VAT) refunds, and what about capital gains tax (CGT)?

A Building houses to sell on is quite normal but it usually happens in a less formal, more gradual way. You are setting out with a commercial approach and this is quite different from self-building. A self-build mortgage applies to people who are using the house as their main residence. Your case is likely to be treated differently. For example, a commercial repayment rate (which is usually higher) rather than residential rate will apply. You will probably have to pay a larger deposit as well. You cannot claim back the VAT if you are building to sell

on. Also, CGT would be payable on the increased value of the houses at the time of sale, if they were not your principal residences during the build. It might be better for the three of you to act as individual self-builders and develop a house each instead.

EURO MORTGAGES

Q Is there a company that provides French (Euro) self-build mortgages?

A Conti Financial Services provides Euro mortgages in France. They will arrange a mortgage as long as the work is carried out by a French-registered builder. An alternative could be to re-mortgage your house in the UK to raise capital to build in France and then apply for a Euro re-mortgage when the new house is complete and reduce the borrowing on the UK property.

INSURANCE AND WARRANTIES

Q We're just about to start the foundations of our new build. The timber-frame company we're using is erecting the house to shell stage. They are working with an independent structural engineer who has his own insurance, but which is not linked to their (or any) limited company. They say that his certificates will mean we don't need any other warranty. Apparently his insurance has no time limit. Is this going to be accepted by potential purchasers' lenders if we decide to sell in the future? This route saves us nearly £31,000 but we don't want to make the wrong decision.

A Ask the structural engineer for a copy of his policy to confirm that the above is correct. On the other hand, it may only cost you about £1,000 for a Zurich policy independent of his cover. Contact Zurich for more information about their cover.

INITIAL DELAYS

Q I'm in a village self-build scheme and experiencing problems with site, public-liability and general insurance cover. I've had a quote for £20,000 for a year of cover, which seems excessive. Also, 40 per cent of the scheme is being funded by the Housing Corporation but they are currently reviewing their criteria, and this is delaying our start. We don't seem to have made much progress after six years, and although the plans were finally approved five months ago, the architect has yet to produce any working drawings. Is all of this slightly excessive for twelve houses? I would value your opinion.

A The quote you have received for the insurance seems high, so contact a few other providers for comparative prices. The only way to make progress with the Housing Corporation and your architect is to be in constant contact with them. Ask what is causing the delays and agree a realistic timetable for the future. Talking to the local councillor about your problems may make the Housing Corporation speed up. Take notes of all meetings and telephone conversations and let the other parties know you are doing so. Write a diary of events to date and give a copy to all parties – this will show them that you mean business. Above all, be persistent and demand regular updates on progress.

AVOIDING CAPITAL GAINS TAX

Q I am having a bungalow built. When completed, I will move in, but I plan to sell it within 12 months. Will I incur CGT? If so, how can it be avoided?

A There is no time limit on how long you must stay in a house to avoid paying CGT. The only requirement is that the house must be your principal residence. Contact the Inland Revenue for a copy of *An Introduction to Capital Gains Tax (CGT1)*. Go to www.inlandrevenue.gov.uk/cgt for further information.

UNDERSTANDING CAPITAL GAINS TAX

Q I live in a house that is divided into six flats (each of them owner occupied) and we collectively own the freehold. The house is in a poor state of repair and we are considering selling to a developer and then self building. We have already had a design drawn up but wonder about the CGT situation. Our intention is to demolish the existing house and build ten new flats on the site. We would retain a flat each and sell the remaining four to repay the loan we would need to cover the construction costs.

There would probably not be any profit involved, but would we be liable for CGT?

A CGT is levied where the intention is to build for profit and/or the property involved is not the main place of residence. It is even levied if the extra flats you build are given away. The Inland Revenue has a good website on this subject (see page 241) where there is a wealth of information to help you. Visit www.inlandrevenue.gov.uk/cgt for more details.

GARDEN PLOTS

Q We are considering purchasing a house with a large garden and building in the grounds. What are the tax implications when we sell the original property and move into the new house, or sell the new house and remain in the original?

A Capital gains tax (CGT) is not due provided that the house you are selling is your principal residence at the time of the sale.

TAXING ISSUE

Q When does council tax have to be paid on a self-build project?

A Council tax becomes due when the property is completed and has been inspected by the district valuer. He will then place it in the appropriate band and the tax is then payable even if the property is not occupied.

COUNCIL TAX

Q We've just finished our house and have been rated in the top band for council tax. Apparently, in our area, the valuation band is based on the market value of the house as of 1 April 1991. What I would like to know in order to challenge the evaluation is how much property prices in the Northwest (specifically Formby, near Liverpool) have gone up since 1991 and also, how much building costs have increased in the same period.

A Telephone the Royal Institution of Chartered Surveyors whose information service may be able to help. Failing that, try your local estate agents. Based on an index of 100 in 1976, building costs increased from 355 in 1991 to 508 in 2001. The provisional figure for 2003 was 567 and the forecasted figure for 2004 is 593.

COMMUNITY CHARGE

Q I'm making a planning application for my new home, but have been told by the council that I'll have to pay a 'community charge' of over £1,000 before permission can be granted. Can this be right?

A Councils can ask for sums of money to be paid when planning permission is granted for a new house. Known as commuted payments, they are usually only asked for on schemes of more than one house. They usually relate to the provision of local services, such as schools, play space, libraries, and to local road or parking schemes. Sometimes the amount to be paid relates to the number of houses, and sometimes to the number of bedrooms in those houses. In your case, it's unusual that the council are charging it on a single dwelling. If you're making a planning application for a new house, including a conversion, it's best to check early on if the council asks for any commuted payments.

SUBSIDENCE

Q I'm buying a derelict farmhouse and barn that have suffered subsidence owing to old mine workings. An application has been submitted to rebuild the buildings, but would it be possible to rebuild them elsewhere on the land, away from the mine workings?

A There are two points to look into here. The first is whether the farmhouse can be rebuilt as a dwelling at all. You describe the farmhouse as derelict, and it's possible the council could decide the residential use of the building has been abandoned if the structure is too far gone. If it's been used for another purpose, such as farm storage, or if it's been empty for a long period, find out what the council thinks. Assuming the council is happy with the general idea of rebuilding, you can broach the subject of relocation. This would be a replacement dwelling, and most councils have specific planning policies governing replacements. Many expect the new house to be on, or close to, the footprint of the old. Much, though, depends on whether any new location offers advantages over the old. Moving to a less prominent location might be regarded as a benefit that would outweigh any technical conflict with policy.

DRAWINGS

Q Where can I obtain drawings to submit for outline planning permission for our proposed house and how much will this cost?

A Look in local business directories for an architectural technician. He or she will prepare your drawings for an average fee of £15 per hour. Alternatively you could go to a firm of architects who will charge about 8 per cent of the construction cost for the drawings and obtaining permissions but probably less for obtaining outline planning permission only.

LAPSED PLANNING PERMISSION

Q We've purchased a cottage with an orchard and garden. In the deeds we found an outline planning permission for two detached dwellings, one of which was built in 1991. We approached the area planning officer who inferred that planning permission for the other house had lapsed. Could you clarify this? Didn't building one house constitute a start on the whole development and couldn't we proceed with the second house as a completion of the original permission?

A If both buildings were included in one permission, construction of one constitutes implementation of the permission as a whole and the full development, that is, the second building, can be built. If, however, design and layout details of the second house weren't approved along with the first one, you probably wouldn't be able to continue to build under the original permission. Double-check the permission to make sure both buildings were included in the one consent, with all details approved and conditions attached to the permission satisfied.

CHANGE IN PLANNING

Q We have bought a bungalow situated in half an acre of land. It is surrounded by large houses that were built six or seven years ago. Before buying the bungalow I rang the planners who told me it shouldn't be a problem to make it into a house, especially as we are surrounded by them. After buying, we decided to replace the bungalow but the planners told us the law had changed and this wouldn't be possible. The only alteration we could make was to put a dormer in the garage roof. Could you please advise?

A It sounds as if you may have a case worth looking into a little further. Your best bet would be to get a professional assessment of your application from a planning consultant to see if it stands a good chance of success at appeal, or whether a revised application might be more appropriate. Without knowing the full facts and lie of the land it's impossible to say whether you have a good case. Such decisions are difficult and there is, presumably, quite a bit at stake, which is why you're better off taking advice from someone familiar with the issues and processes. If the prospects of an appeal are reasonable, give it a go. It's really worth taking the time to do this properly because you only get one chance with an appeal. It's unlikely that a change in the law has made a significant difference in this instance. However, it's very possible that there's been a change in council planning policy since you bought the property. This does happen and can render opinions given in good faith by planning officers at the time of the original enquiry, out of date.

LOSS OF BUSINESS SPACE

Q I'm hoping to build on some land that's been used in the past for storage. It's an ideal plot in a residential area. The council, though, says it doesn't want to lose the business use of the site and wants to see it advertised for sale for a year to make sure there's no demand for it as it stands. Is there a way round this?

A Councils often have policies that resist the loss of business space, which probably explains their stance here. That said, a storage use in a residential area sounds at best unattractive, and possibly generates a lot of traffic. Neighbours would probably be only too pleased to have a house instead — it might even improve the value of their own properties. Talk to neighbours and, if they're supportive, talk to your parish or town council and your local councillor from the planning authority. Strong local support and visual and traffic improvements may be enough to override a policy-based objection from the planners. If you have plenty of support, make a planning application and ensure supporters write

in. Even if the planning officers are opposed, you might be successful if your application can be presented to the planning committee.

HALF A SITE

Q I recently applied for planning for a bungalow in my garden. I was told that the site was outside the local development area and would be refused. I went to the council offices to view the local plan and it seems that when the boundary for the development area was drawn, it cut my property in half. Could you advise me as to the best course of action and the cost?

A Your best chance of getting planning permission is likely to be through the local plan process. If the site of your bungalow is outside the council's boundary, the council will probably refuse it. If there are any special circumstances you can put forward, an appeal might succeed. However, trying to get the boundary changed into a more sensible position, and one with more relevance on the ground, would be the best route. The drawback is the time this takes: plans are reviewed at 5–10 year intervals and the process takes several years. Things are further complicated because the present system of local plans is due to change. The bill is going through Parliament at the moment and is yet to be finalized. The cost of pursuing an amendment to a plan varies, depending on how you choose to support your objection. To give you some idea, the range is likely to be from £750 to £2,500.

CONVERSION IN A GREEN BELT

Q My family has a barn that has potential for conversion. Some recent advice suggests that the rules on conversion and/or development within a green belt area are changing in a couple of years' time. Where can I find more information on these changes and their implications?

A Green belt restrictions are in place to stop new builds popping up in the countryside. They do not prohibit barn conversions, and green belt policy specifically encourages the reuse of rural buildings. However, general policies favour conversion to commercial use over residential, which could be a bigger stumbling block for your proposal. Any forthcoming changes in legislation are unlikely to have a significant impact on conversions, although, naturally, these things tend to get harder over time.

OCCUPANCY LIMITATION

Q We are nearing completion of a self-build house in the centre of a village. Towards the rear of our plot is a mobile home, which was granted planning permission 20 years ago with a condition limiting occupation to a named user (now deceased). What is the status of the mobile home now? What are the chances of getting the condition removed? Could an application to remove the condition damage its current planning status?

A The status of the mobile home depends on how it has been used. For example, has anyone other than the named person occupied it at any time? If not, it is probably lawful to retain the home but not for anyone to occupy it. Getting the occupancy condition removed could be possible and it would be worth looking back at the council's planning records to find out why a personal permission was given in the first place: there might have been extenuating circumstances. If the home has been occupied for many years without problems, there seems little reason for not allowing this to continue.

HOLIDAY LET TO HOME

Q I want to convert a farm building currently used as a holiday let into a family home, but council planners tell me they will not allow a change to private residential use. Another building has been converted to a house nearby, so can I therefore challenge this stance?

A The advice the council has given you reflects both government planning policy guidance and the council's local plan policies. Generally, to obtain permission for residential use, you have to find special circumstances demonstrating that the holiday use is no longer viable: perhaps access is poor, the building is in a remote location, or there is a general lack of demand. This usually has to be proved by a 6–12-month market exercise. Accordingly, where permission has been granted for a holiday let use it may be

difficult, if not impossible, to revert to a full residential use. However, look up the planning file for the similar case property you mentioned at the council's planning department, to find out how planning was achieved. It could be that the site in question benefited from older, more helpful policies than apply now, as the move against residential conversions has been slow to filter down from government to local planning (introduced in 1997). If in doubt, seek advice from a planning consultant.

RESTRICTIVE COVENANTS

Q I purchased a barn for conversion from my local authority 23 years ago. The sale included ⅔ acre of land with a restrictive covenant. I now wish to build a house on half of the land that is affected by the covenant. The local authority is asking for half of the anticipated development gain. I wonder if there is there a recognized formula for valuing the release of a restricted covenant.

A These matters are normally settled by market forces. Your local authority may ask for 50 per cent of the enhanced value of the land, depending, to some extent, on the price you paid for it 23 years ago. A valuation surveyor will be able to give more advice.

MAXIMIZING POTENTIAL

Q We are wondering whether any sort of permission could be gained on our land. We live in a three-bedroom house (which could do with quite a bit of renovation) set within five acres of unused cider orchard. We are outside our village boundary, so assume planning would not be simple. We have some old brick-and-stone outbuildings, which were once inhabited but are currently used for storage. There's also a very ugly corrugated aluminium chicken broiler unit that is an eyesore from the passing road. Our initial thought was that the chicken house could be removed and the buildings converted to a holiday cottage. However, we're happy to consider any options, even demolishing our current home to build a new house.

A Your holiday cottage idea is a good one, because councils are keen to allow the reuse of rural buildings for business purposes. Holiday cottages count as a suitable rural business: they create employment and bring money-spending visitors into rural areas. Provided the conversion can be undertaken without too much in the way of radical alteration or extension, the scheme should be acceptable. Chances are, your council has a specific policy in its local plan that sets out the criteria such schemes must meet. Have a look at the plan at the council offices or local library and see how your project matches up. Do not despair if you can't meet all the criteria: the removal of the chicken unit is a bonus that could be used to outweigh any perceived shortcomings. Demolition of your existing house is also an option and, again, your council's local plan should have a policy setting out criteria for replacement dwellings in rural areas. Such policies usually specify that the new house shouldn't be significantly bigger than the existing one and should be located in more or less the same position. These policies can be flexible and are often vaguely written. Talk to a planning officer to see how the policy is usually interpreted locally. Bear in mind the broiler unit could be brought into the equation to justify some additional space in the house. The broiler unit is unlikely to buy you an exact equivalent amount of space, but it should certainly feature as part of the package of enhancements that your scheme will bring. There is the possibility of converting the outbuildings to a new house, but the chances are slim if there's obvious scope for business use.

LIMITED EXTENSION

Q My wife and I have recently had an offer accepted on a small bungalow, which sits in a rural third of an acre. We have often read that people double the size of their homes through extension but have been informed that we cannot increase by any more than 30 to 40 per cent of the current size. Why is this?

A Many councils place limits on how big extensions can be to houses in the countryside, and they vary as to how generous they are: some allow 50 to 60 per cent, others only 25 to 30 per cent or, sometimes, specify a maximum floor area. Look up the relevant council planning policy in the local plan, which you will

find at the council offices or at a local library. Read the policy and accompanying guidance carefully. Are limitations specific or couched in more general terms? If the latter, it is, perhaps, a question of interpretation. The intentions of such restrictions are to limit the amount of building that takes place in the countryside and to preserve the stock of smaller rural dwellings so that lower-income or smaller households are not excluded from living in the countryside. If the policy is very clearly against you, it could be difficult to overcome but you should investigate the officer's decision nevertheless.

GARDEN CONVERSION

Q I understand that it is very difficult to convert agricultural land to garden, but is this made any easier if the land has actually been used as a garden for at least 50 years? It is mainly laid to lawn with flowerbeds and fruit trees, is about half an acre in size, and abuts my house, which is in the green belt. I do not own the garden but may have the opportunity to purchase it at a later date.

A You are quite right that it is difficult to convert agricultural land to garden, because councils tend to view it as an 'erosion of the countryside'. Many have policies aimed specifically at preventing it, which are usually more restrictive in green belt areas. Luckily, a change of use becomes lawful when it has existed for 10 years or more – so you should be fine.

OUTBUILDING CONVERSION

Q Immediately behind our house we have a two-storey garage, store and hayloft. It is approximately 300 years old and was partly rebuilt in 1987. Prior to rebuilding, the store had always been used as a house and we were wondering if planning might be granted to convert it back to a house and to move it. Also, how big would we be able to build, as it is small?

A I imagine your property is in the countryside rather than in a village or town. If that's the case, your plans will be dependent on the council's conversion policy. Whatever the building was, it seems it is now an outbuilding related to the main house. The fact that it was once used as a house isn't completely irrelevant, but it's not a decisive factor either. Find out what your council's conversion policy is and see how this building would measure up. Generally, you are not allowed to relocate or significantly increase the size of conversion properties.

You'll also need to consider practical points, such as the relationship between your house and the outbuilding. Consider things like whether each property would have privacy and whether access to one house would have to pass close to the other. If there are no practical problems, and you comply with the council's rural buildings conversion policy, then you're in with a chance and it'll be worth speaking to a planning officer.

SELF-BUILD SAVINGS

Q With careful planning and budgeting, how much could be saved in percentage terms by building your own house compared to buying one?

A If you physically do a lot of the building work yourself you could save as much as 30 to 40 per cent. If you choose just to manage the build, you should produce savings of about 15 to 20 per cent.

DOING THE DONKEY WORK

Q We have planning permission for two 130 m² bungalows in the rear of our garden and would like advice as to how much a contractor would charge to build them per square metre. Also, how much could we save if we did a lot of the donkey work ourselves?

A Many factors influence building costs but, for budget purposes, work on £750 per square metre. This equates to £196,000 in your case, plus about £25,000 for a small access road, drains, paths, fencing and landscaping. This figure could be reduced by 10 to 15 per cent if you managed the site yourself and probably even further if you used a timber-frame firm to erect the shell for you.

WHAT DOES 'A SQUARE METRE' INCLUDE?

Q I am planning a two-storey extension at the back of my house, including a bedroom in the new loft space. When one talks about building costs as a £ x per square metre, does that also include the cost of the roof construction, or should I budget separately for that? What figure per square metre should I use?

A Square metre prices include all costs from the foundations to the roof, but exclude external works such as drains, paths, drives and so on, and should be calculated on measurements taken between external walls. In working out the approximate cost, much depends on the quality of the specification you want, but for your budget work on £700 to £750 per square metre plus £8,000–10,000 for drains, paths, drives and fencing.

PROJECT COSTS

Q I want to build a one-and-a-half storey home of about 150 m². The ground reports call for standard strip foundations and say that the site is level. What can I expect to pay a contractor for a full build and how much would it cost to employ a project coordinator?

A Work on £750 per square metre for budget purposes and add £8,000–10,000 for paths, fencing, drains and so on. Expect to pay between £15 and £20 per hour for the services of a project manager. A manager will not need to be on site full time, so his or her fees should equate to about £3,000 for a standard build.

AVERAGE PRICES

Q I am looking to build a three-bedroom bungalow of about 120 m² using traditional construction. I have seen various figures published for costs per square metre ranging from £500 to £1,700. I am looking for the cost to build the shell only. Can you help me with any figures?

A The current cost of building a house is about £750 per square metre and approximately 50 per cent of that would cover the cost of the shell, that is, the foundations, external walls, windows, external doors and roof. In your case, that would be approximately £42,000. Add a further £8,000 for drains, paths, drives and fencing.

SHELL PRICES

Q Could you give me an idea for building just the shell of a house – that is, the foundations, roof and external walls. Also, how much should we pay for architect's services?

A Work on £300 per square metre for constructing the foundations and shell of your proposed house. Architects will probably charge about eight per cent of the cost of the construction work for the design, obtaining planning permission and appointing a builder (pre-contract work) and three to four per cent for looking after the job once it starts (post-contract work). These fees are not fixed, so you should shop around for the best deal.

HOUSE MODELS

Q If I have to build a model-size house, what are the best materials to use and where can I get information on how much everything would cost if I were to build the house in real life?

A A model can be made from stiff white card and balsa wood. If you base it on one of the projects featured in a self-build magazine you will find detailed information on how the build was achieved and an analysis of costs.

BUILD BUDGET

Q I am at an early stage in a self-build and am trying to assess how much the project will cost. I have a budget of about £95,000 for the land and construction of a four-bedroom house in the 100 to 130 m² range. Is this possible? I expect to pay £30,000 to £40,000 on the land, leaving £55,000 to £65,000 to build the house.

A It may be just possible to build a house for £60,000, but your budget will mean you have to be incredibly careful with your costings. Your build figure equates to £550 per square metre for 100 m² and £430 per square metre for 130 m². Self builders who have managed such budgets have usually carried out a significant part of the construction work themselves. Also, your land budget means you can afford only to build in a limited area (see the Land Directory for affordable locations).

QUOTE, MISQUOTE

Q We are hoping to start a home-extension project shortly. We have permission to erect a single-storey rear extension (163 m²) and a two-storey section to the side of the property. After reading a recent magazine article, we calculated that the build cost would be about £146,000. Although a firm of engineers/designers produced a similar figure, we are only receiving build quotations in the region of £250,000, which is far beyond our budget. How can we get a fair quote?

A Estimated costs of construction work and what you are asked to pay can vary widely. Your figure is possibly correct but it is hard to make allowances for some contractors (for example, if they have full order books they are probably not worried about having the job). My advice is to persevere with the enquiry process until you find a builder who will give you a more acceptable quotation. The temptation is to accept the present lowest quotation and negotiate to reduce the scope of work or standard of specification until it matches your budget figure. However, I would not advise doing this, because you will not get value for money.

DEMOLITION COSTS

Q I have seen a decommissioned water-pumping station for sale on a very attractive site. The purchaser is responsible for demolition, excavation and locating the incoming main. There is also a large cylinder that needs to be removed. How much should this work cost?

A Demolition costs vary widely depending upon the type of building involved and particularly whether there is any salvage value in the materials. But, work on £7 per cubic metre for the demolition and excavation. You may find a local farmer or developer who needs a hole filling in and they may pay you for the rubble. The cylinder may have scrap value and you might be able to sell it on (unless it once held fuel, in which case there could be complications with the Health and Safety Executive about its disposal). Obtain quotes before you make your offer because there are too many unknowns to take a chance on.

RENOVATION COSTS

Q A three-bedroom house is for sale for £85,000 and needs to be gutted totally. Is there a way to assess the costs involved?

A A rough-and-ready method is to calculate the costs as a percentage of a new-build project, based on £750 per square metre. First, apply this figure to the whole area of your house, then work out how much of the building needs work. For example, you might estimate that 50 per cent of the house needs work, in which case you would halve your new-build figure (£350 per square metre). Based on a typical three-bed semi with an area of about 110 m^2, therefore, the cost would be 110 x £350 = £38,500. Alternatively, you may wish to employ a quantity surveyor to prepare a cost estimate for you. He or she would charge you about 1.5 per cent of the value of the building work and the estimate would be a useful tool in the running of the job.

UNDERPINNING

Q We are negotiating to buy a bungalow that needs the foundations underpinning. Could you give us an approximate cost for this type of work?

A Expect to pay about £400 to £500 per linear metre for underpinning a normal house wall. This figure could increase if ground conditions are poor and there is a need to go deeper than 1 m. You will need the advice of a structural engineer or a firm specializing in underpinning.

TRENCH-FILLED FOUNDATIONS

Q I would like an indication as to what it will cost to build concrete trench-filled foundations for our new home.

A You should expect to pay between £50 and £70 per metre for trench-filled foundations.

EXTENSION COSTS

Q How much would it cost to add a second floor and build an extension to a two-bedroom bungalow?

A The cost of a loft conversion varies in direct proportion to the size of the space being converted, and ranges from about £550 per square metre for a loft sized 20 m^2 to £450 for one sized 30 m^2. A greater demand for this kind of work in recent years means that builders are able to charge higher fees, so you must shop around. A single-storey extension with a flat roof will range from £1,200 per square metre for a 15 m^2 extension to about £650 for a 24 m^2 extension.

BUNGALOW CONVERSION

Q I want to convert my bungalow into a house by adding a second storey and enlarging the footprint slightly. I also want to use around 30 m^2 of the current footprint as an integral garage. The bungalow is 143 m^2. The build would involve adding around 36 m^2 at ground level and then adding the second floor directly above this of around 179 m^2. What should I budget for build costs?

A Expect to pay about £20,000 for the extra accommodation at ground level. Taking off and re-fixing the roof (using the same materials) should cost about £15,000 and the extra first floor should be about £90,000. This seems a lot of money for what you would achieve, but talk to some local builders and try to get a rough price from them before committing yourself.

GARAGE FIGURES

Q Can you give me some ballpark figures for building a double garage with a pitched roof and a room into the roof space for use as a guest bedroom?

A Expect to pay about £25,000 for a double garage with a bedroom in the roof space (with two dormer windows and a staircase). Add a further £3,000 for plumbing if you install a bathroom. The garage alone should cost around £14,000.

A NEW GARAGE WITH LIVING SPACE

Q We would like to build a double garage with a room over the top. It should have access to our house on both levels and good acoustic insulation. How much should this cost and how long should it take to build?

A A contractor will probably charge about £45,000 including the connections to the existing house and extra insulation. You could probably save £10,000 by doing some of the work yourselves. The project should take between 12 and 14 weeks. Make sure you get at least three quotes from contractors before you commit.

COSTING A NEW STOREY

Q Could you give me an approximate cost for converting a single-storey stone building into a two-floored home of 3,000 m^2?

A If the external walls and roof are in good condition allow £550 per square metre for each floor. This figure will vary according to the quality of specification you are working to and how much of the work you will do yourself.

CONNECTING SERVICES

Q I have just bought a building that has been vacant since the 1950s. It is stone built with a slate roof that looks in fairly good order. There are no electricity or water supplies but connections are nearby. What sort of price can I expect to pay to connect to these services?

A Costing the supply of services is difficult because the utility companies have a near-monopoly on their products and are prone to take advantage of this. However, under normal self-build circumstances, where the existing services are adjacent to the plot, costs will be around around £5,000. The only way to find out is to ask the providers themselves because, unfortunately, they don't work on standard charges.

COST OF SERVICE CONNECTIONS

Q I am preparing a budget for our proposed bungalow. I've allowed £5,000 for service connections. Is this enough?

A Each authority is different but they seem to have one thing in common: a reluctance to provide cost information that isn't qualified. However, when comparing with other self-builds, it would appear that £5,000 should be adequate.

POWER CABLES

Q We are considering restoring an old farmhouse in a remote rural area. Could you give us an indication of the costs involved in laying power and telephone cables in the same trench?

A The cost of the services, including provision and laying cables, will depend on the relevant service providers, so go to them direct for estimates. The cost of digging a trench 750 mm wide and 500 mm deep (including backfilling) is about £9 per linear metre.

COSTING TROUBLES

Q I am preparing a build budget for a timber-framed house but am struggling with costs for underfloor heating, electrics and plumbing. The house will be approximately 325 m². Can you help?

A The cost of electrical and plumbing work can vary widely, mainly owing to the quality of the fittings used. For budget purposes, work on £20 to £25 per square metre for underfloor heating, £3,000 to £4,000 for electrics and about £7,000 for plumbing. The best way to obtain more accurate figures is to get quotes from contractors.

PLUMBING AND HEATING

Q I'm extending and renovating my current home to make five bedrooms, two en-suite bathrooms, one main bathroom, a large kitchen, a large living/family room, a study and a downstairs toilet. I also want a new heating system. I've had a quote for £18,000 from a plumber including a £4,000 allowance for bathroom suites. The quote allows for 17 radiators and three towel rails. How can I work out if these figures are reasonable?

A Allow £250 per room for labour, £50 per radiator, £1,000 for a decent boiler, £1,000 for a pressure cylinder, £350 for pipe work, £250 for waste pipes and £300 for controls. You'll also have to add VAT to these figures.

PLASTERING COSTS

Q Could you please give me an idea of plastering costs? I have a small stable unit that has just been plasterboarded and have received a quote that seems very high. Is there an average hourly rate or price per square metre?

A For two-coat plastering work you should expect to pay around £10 to £12 per square metre, allowing £9 for plasterboard and £7 for board finish.

STREAM UNDER HOUSE

Q We need to establish whether an underground stream lies beneath the site we're proposing to build on. If this is the case, should we proceed? Could it affect the development cost and how could we work around the problem?

A The stream problem is tricky. You need to establish how far below the surface it runs. If it is not too deep it could be exposed and either diverted or made to run through a concrete culvert. If it is very deep, it should not be a problem but you should consult a civil engineer for advice.

DRY-STONE WALLS
Q How much should I expect to pay for labour and stone in constructing a dry-stone wall? The wall will be 1.5 m high and 100 m in length?

A Contact the Dry Stone Walling Association. They will be able to give you an idea of the costs and may be able to recommend a wall-builder in your area.

ACCESS ROADS
Q We are attempting to buy a plot of land but the sale is dependent on the cost of putting in an access road. Can you tell us how much a basic track should cost to build. It needs to be suitable for a twelve-ton delivery lorry?

A For a temporary access road allow about £14 per square metre. If you want a permanent tarmac, work on £35 per square metre. Both of these figures exclude drainage costs.

Construction

CALCULATING AREA
Q How do I calculate the size of my home? Do figures that appear in most self-build magazines include areas covered by landings, stairs and walls?

A You should measure from the inside of the external walls across stair openings, internal walls and chimneybreasts for each floor of the house. In the case of habitable rooms in the roof space, the usable floor area is measured excluding the dead area at the angle of the roof and floor.

ENOUGH LAND?
Q I have a piece of land measuring 6.87 x 5 m. Is this large enough to build a single-storey mews-type house – either studio or one-bedroom?

A Assuming you could use all the land for the new building, you could fit a studio-type flat into this area. The question is whether you would get planning permission for such a small building. Getting permission depends on the scheme meeting the council's criteria for new houses. First, is the land in an area where permission would normally be granted, such as a town or built-up area of a village? Would it fit into the general pattern of development? Would the new house enjoy adequate levels of light and privacy, or cause any such problems for adjoining properties? What about parking? Very small houses do sometimes get permission, but much depends on the individual circumstances of the plot in question. Good design is often the key and your first step might be to talk to an architect or planning consultant who can assess the prospects.

EXTENSION FOR A TIMBER-FRAMED HOUSE
Q I am considering extending my home. The property is timber-framed with a brick outer skin. Is it possible to build a timber-framed extension or would a brick-and-block extension be more suitable?

A You are free to construct in either timber frame or brick and block, as long as you get supporting structural calculations from an engineer and the extension adheres to Building Regulations. It would be helpful to have a set of the original plans towards the preparation of the new layouts and construction drawings of the proposed works. If you haven't, you may be able to get a copy from the manufacturer or your local planning office.

ORDER OF WORK
Q We are moving to a house where a two-storey extension has been built but left as a shell for us to complete. We need to plaster, build partition walls and install the electrics, plumbing, carpentry and so on. The extension has two bathrooms and a kitchen. In what order do we need to organize the work? (The upstairs has chipboard flooring laid but not fixed to give our electrician and plumber access.)

A First, a joiner needs to fix the floor and the wall partitions in timber studs. Next, an electrician and plumber need to first-fix pipes and cables.

Insulation needs to be fixed next, before the plasterboard and skimming. The joiner can then return to second-fix the doors, architrave, skirting and so on. Then the plumber and electrician can second-fix radiators and switch boxes.

FOUNDATIONS AND SERVICES

Q When constructing a trench-fill foundation (when the level of the concrete finishes just below the surrounding ground level) what is the best way to get drains and services to the inside of the building? Can ducts be laid across the trench before pouring the concrete? Also, won't the top of the concrete be in the way of the gullies that need to be close to the outer wall?

A When the trench foundations are constructed allow for 45 mm to be added to the damp-proof-course (DPC) level. Also, place 150 mm ducts in the trenches to allow for the location of 100 mm drains later.

TREE IMPLICATIONS

Q We are about to buy a plot but a large tree on the land is protected by a preservation order. We are told this means we have to use piled foundations to protect the tree roots. How much would this cost?

A You need to confirm with a civil engineer that piled foundations are necessary. If they are required, the depth of the piling will depend on the nature of the ground. Also investigate whether it would be necessary to pile all of the foundations or just those adjacent to the tree. This may depend on the species of the tree and whether the roots are growing mainly vertically or horizontally. When taking into account all of the above variables it is difficult to calculate exact costs, but you should expect to pay about £50 per linear metre per pile. Take professional advice on the design of the piles.

WHAT ARE TRENCH BLOCKS?

Q I am undertaking a self-build project and have got as far as pouring the concrete for the strip foundations. I have decided to use timber frame for the construction of the house and am using 140 mm timbers. For the footings, I have been told that I can either lay a 100 mm block with a 50 mm cavity and then a 150 mm block inside; or lay a trench block. I do not know anything about trench blocks except the fact that they are lightweight and cost more than concrete blocks, but are easier to lay. My question is to what dimensions are these blocks made? How much more would I be likely to pay per block. And are there any other ways of constructing the footings?

A Trench blocks vary in size from 290 mm to 350 mm wide, dependent on where they are made. Costs vary in accordance with the distance from the manufacturer to your plot – your merchant will quote for all three types of blocks. The third option you have is to 'mass-fill' the trenches with concrete. This costs more in materials, but less on labour.

CELLAR CONVERSION

Q I'm considering buying a five-bedroomed Victorian terraced house that has three large cellar rooms measuring approximately 56 m². The back cellar room is at ground level and has an access door. The cellars seem dry and I would like to know a rough cost of converting the space into a one-bedroom flat to rent out. I need to install sanitary facilities – is this possible given that the existing soil pipe is at the front of the house and the cellar floor is below ground level at this point?

A You should be able to fit out the cellar for about £300 per square metre, providing there are no structural problems. Ask an architectural technician to consider the drainage problem on site.

PREFABRICATED HOMES

Q I am considering buying a prefabricated home that cannot get a mortgage and which has been designated as defective. Similar properties on the estate have been upgraded by what appears to be a new outer leaf of brick. Is this cladding simply built around the outside or does the existing skin have to be removed first?

A It's probably not cladding, as most prefabs need a full outer skin to create a cavity and meet insulation standards. This also has to be done to meet Building Regulations approval. Go to the local planning office and ask to see the plans for one of the renovated properties on your estate. This way you'll be able to see exactly what they've done, although consider that regulations are probably stricter today.

WALL CAVITIES

Q I am building a two-storey extension to my own house with block walls that have 100 mm cavities. Is there a minimum width cavity? And can it be fully filled with insulation – assuming the correct U-value is reached? If I have to do a partial fill, is there a minimum residual cavity width?

A The cavity can be fully filled at 50 mm and up to 100 mm. Or, the cavity can be 100 mm partially filled, with 50 mm insulation. Whoever drew up your plans for Building Regulations approval must submit calculations for the energy rating, which take into account the number of windows, floor design, roof design and existing structure. Only the combined consideration can determine the cavity width and insulation requirement.

HEATING FUELS

Q Can you advise on the best type of heating system to install in a new build that could be converted to gas when a supply becomes available in our area in two or three years' time.

A Buy the cheapest oil boiler you can find at the time you build. This can be easily replaced with a gas one when the supply becomes available.

NATURAL LIGHT TO HALLWAY

Q My husband and I are planning a lean-to extension to the rear of our property. There is a landing window just where the roof is going to be. If we take it out it will make the hall too dark. Is there a way around this?

A Yes – fit a Velux roof light with a 'shaft' down to the ceiling. Or, fit a Monodraught Sunpipe or one of Solarlighting's Solartubes to the roof which will 'pipe' the light to the hall ceiling. These companies advertise in self-build magazines.

BESPOKE STAIRS

Q I need to build a staircase which requires a turn but am unsure how to achieve this. Can you give me any advice?

A Find a local joiner. He or she should be able to design a staircase for you and supply materials.

GAP BELOW ROLLED STEEL JOIST (RSJ)

Q I have just added an extension to my home, but unfortunately the underside of the main RSJ is too high for the new internal doors. There is a deficit of 10 cm to make up. It was my intention to pack out the underside of the RSJ with timber. I don't trust any adhesives to secure this packing and wanted to know if it is possible to drill and tap the bottom of the RSJ and bolt the piece of timber up to it?

A There are two solutions to this problem: you could either drill and tap the beam to fix the timber, or you could just drill the beam and secure the timber with a coach bolt. The drill-and-bolt option is the simplest.

ROOFING COSTS

Q What is the going rate for laying slate roofing?

A The cost of laying slates depends upon the size of the slate being laid. Large slates are cheaper per square metre because each one covers a larger area. As a guide here are some approximate costs for different size slates. These figures also include laying the felt and battens:

Slate size	Labour cost per m^2
405 x 205 mm	£16
610 x 305 mm	£12
510 x 255 mm	£14

If the roof is complicated in terms of extra valleys and other abutments and junctions, add 10 to 15 per cent to these figures.

RAFTERS OR TRUSSES?

Q The drawings for my house show a roof constructed of purlins and rafters, but my joiner says trusses would be a lot easier for him. Is there any reason why I couldn't use trusses instead?

A Your joiner is right in saying it will be easier and quicker to fix trusses rather than purlins and rafters. However, your drawings provide an engineered solution approved by Building Control, and your architect (or whoever drew up your plans) should be consulted before any changes are made. Truss suppliers are able to design to meet the structural requirements and will provide the calculations to prove the integrity of the trusses to Building Control. You should also investigate any cost implications before acting on your joiner's advice.

RENDERING A DORMER

Q I want to render the gable and cheeks of a roof dormer. The dormer is built with timber and has one layer of 18 mm ply externally. What preparation is needed before applying the render and can you recommend a render product?

A Try a product called Rendalath – it should meet your needs. A 25 mm air cavity between the timber and the product will be necessary. Rendalath is available from BRC building products.

SOUNDPROOFING

Q I live in an apartment and have a noise problem from the flat above. I have read about soundproofing underlay and ceilings. Can you advise what type of building professional I should ask for advice.

A You should consult a building surveyor or architectural technician. Both are listed in your local business telephone directory under those specific headings. Describe the requirement over the telephone, assess the insight and first-hand knowledge expressed in the response and the price they quote for a site survey and recommendation, which you can then pass on to a builder for a fixed-price quote.

HEIGHT OF POWER

Q How far should electric sockets be from the floor?

A Building regulations require sockets to be minimum 450 mm from the floor and a maximum of 1200 mm.

LOFT FLOOR

Q I have the potential to convert the roof space in my bungalow, but I don't know how to install a floor for the conversion. Can you advise me?

A You can't convert a roof space to a loft without Building Regulations approval (and, perhaps, planning permission) and you can't apply for regulation approval without structural drawings.

Have the roof surveyed by a local architectural technician, who can determine the structures. Only that will tell you what you need.

NOISE REDUCTION

Q What materials can be used in the construction of a proposed new bungalow to reduce noise from the nearby raised dual carriageway? Can you advise me?

A There are a few things you could try: install triple glazing instead of double; use 140 mm blockwork for the inner walls, instead of 100 mm or a timber frame; and get a quote from a local fencing contractor for a double-skin acoustic fence.

BUILDING A RETAINING WALL

Q I need to build a retaining wall, 1 x 6 m. How thick should the wall be, and how deep should the foundation buttresses be? What materials should I use?

A The foundation should be 300 mm deep and 900 mm wide. You should fix steel L-shaped reinforcing rods into the foundation to the front edge. This will allow a structural 'toe' at the back of the wall to act as a counterbalance. When the foundation is cast, lay 225-mm hollow concrete blocks over the reinforcing rods and a lateral reinforcing bar in each course of blocks, with ties projecting from the front face. When complete, fill the blocks with concrete, which will tie the reinforcing rods into the overall structure. Then 'face' the wall with bricks or stone. If this wall is subject to Building Regulations approval, you will need to show a drawing and structural calculations.

TREE DAMAGE

Q I want to build a driveway between two trees, but have had planning turned down on the basis that the tree roots may be damaged. The trees are Scots pine. Is there any method of building a driveway that wouldn't require proper foundations, so the trees would remain unaffected? The span would need to be around 5 m wide and cross a drainage ditch.

A You could build a bridge over the roots, but this will be expensive. Have a look at the way farmers bridge ditches to drive tractors into their fields. It depends on whether Building Control insist on calculations and regulations.

ALTERNATIVE WATER SUPPLY

Q I have a plot of land on which the only mains water is supplied by the local council and is limited to quantity and time of day. Is there an alternative to this situation?

A You might look into sinking a borehole to establish a well of your own. The Environment Agency should help (see page 242).

Project management

SHOULD I BUILD IT MYSELF?

Q I have no building experience, but would like to manage a self-build on a limited budget. Should I: get a timber frame supplied and erected by a package company to a watertight stage then manage the rest of the work myself; get a timber-frame company to carry out a full build; undertake the full build myself?

A It is difficult to know which is the best course for you as everyone is different. Financially, I would go for the options in the order in which you list them. The second option is the safest because you know how much it will cost before you start, but it is also probably the most expensive. If you are able to identify different aspects of the work (in this case plumbing, joinery and so on), put a cost against them and monitor events carefully. The first may suit you better, but you would have to live with the uncertainty of not being sure of the total cost until a fair way into the project.

ABSOLUTE BEGINNERS

Q My husband and I are interested in building a new home for our family but neither of us has building experience. Would we need someone to oversee the whole project for us and would this mean that we'd see no profit from the build? The land-buying part seems to be the most difficult, and we are anxious that our inexperience could lose us a lot of money before we even start to build. At the moment

we favour using a timber-frame package company. Is this the best way forward for us?

A Thousands of people without any building experience build their own houses every year: the main qualities you need are common sense and determination. It would be extremely unusual for a self-builder not to make a profit on building a house, although the profit can't be realized until the house is sold on. Although, as you say, the most difficult part is finding a plot of land, in normal circumstances it would take extremely poor judgement to lose money on a land purchase deal. Your decision to use a timber-frame company is probably sensible, as it will undertake all of the work, including supervision and project management. Look at the company's designs and find out what services are available, and at what cost. Remember that you are entitled to reclaim the value-added tax (VAT) on materials used in building a new house.

QUANTITY SURVEYORS

Q My architect is submitting plans for a 325 m² house to my local authority on my behalf. I have been very happy with his services, but he seems uncomfortable about managing the project and has advised me to employ a quantity surveyor (QS). What advantage would I gain from using a QS, and what is the cost likely to be? The architect is charging a fixed price to obtain planning permission – is this normal?

A Working on a lump sum to obtain planning permission is normal and acceptable. The sum will be based on your architect's assessment of hours spent, multiplied by his hourly rate. This work is referred to as pre-contract services (including the preparation of working drawings) and is usually set at five to six per cent of the construction costs. If you employed a QS to look after the post-contract work, he or she would handle all of the financial matters, including appointing contractors, making monthly payments, measuring and pricing any variations, settling the final account, and offering any contractual advice that was needed. He or she would charge about three per cent of the construction cost for these services but he or she would not check the quality of the work.

Useful addresses and contacts

The following list of contacts is by no means exhaustive. Most have been referred to in the text and are within their relevant chapter.

Budget and finance

BUILDING COSTS
BuildStore
Tel: 0870 870 9991
www.buildstore.co.uk

www.costofdiy.com

EasyPrice Pro
Tel: 0845 658 9626
www.easypricepro.com

Fast Estimate
Tel: 01764 655331
www.estek.co.uk

HBXL
Tel: 0870 850 2444
www.hbxl.co.uk

PROJECT MANAGEMENT
Insider Project Management
Tel: 01262 605653
www.davidjackson.co.uk

RECLAIMING VALUE-ADDED TAX (VAT)
HM Customs and Excise
Tel: 0845 010 9000
www.hmce.gov.uk

STRUCTURAL WARRANTIES AND SITE INSURANCE
BuildStore
Tel: 0870 870 9991
www.buildstore.co.uk

Self-builder.com
Tel: 0800 0197660
www.selfbuilder.com

Self-Build Zone
Tel: 0845 230 9874
www.selfbuildzone.com

CAPTIAL GAINS TAX (CGT)
Inland Revenue
www.inlandrevenue.gov.uk/cgt

Finding the perfect spot

FINDING LAND
Landbank Services Ltd
Tel: 0118 962 6022
www.landbank.co.uk

Land Registry Office
(England and Wales)
Tel: 020 7917 8888
www.landreg.gov.uk

PlotSearch
Tel: 0870 870 9991
www.plotsearch.co.uk

The Registers of Scotland
Tel: 0845 607 0161
www.ros.gov.uk

Stonepound Books
For books regarding land and planning.
Tel: 01273 843737; 01825 890870
www.stonepound.co.uk

PLANNING PERMISSION
England
Planning Policy Guidance Notes (PPGs)
(being replaced with Planning Policy
Statements)
www.odpm.gov.uk

Northern Ireland
Tel: 028 9041 6700
www.planningni.gov.uk

Republic of Ireland
Ministerial guidelines; Ministerial policy
directives
www.plenala.ie

Scotland
National Planning Policy Guidance;
Technical Advice Notes
Tel: 0131 556 8400
www.scotland.gov.uk

Wales
Planning Guidance
Tel: 029 208 25111
www.wales.gov.uk

Speer Dade Planning Consultants
For professional help with planning.
Tel: 01273 843737; 01825 890870
www.stonepound.co.uk

Design

USING PROFESSIONALS
British Institute of Architectural
Technologists (BIAT)
Tel: 020 7278 2206
www.biat.org.uk

Constructive Individuals
Tel: 020 7515 9299
www.constructiveindividuals.com

Royal Incorporation of Architects in
Scotland (RIAS)
Tel: 0131 229 7545
www.rias.org.uk

Royal Institute of British Architects
(RIBA)
Tel: 020 7580 5533
www.architecture.com

Royal Institution of Chartered Surveyors
(RICS)
Tel: 0870 3331600
www.rics.org.uk

Royal Society of Architects in Wales
(RSAW)
Tel: 029 2087 4753
www.architecture-wales.com

Royal Society of Ulster Architects
(RSUA)
Tel: 0131 229 7545
www.rsua.org.uk

PLANNING YOUR ROOMS
The Building Centre
Tel: 020 7692 4040
www.buildingcentre.co.uk

RIBA Bookshop
Tel: 020 7496 8390
www.ribabookshops.com

HARD LANDSCAPING
Dry Stone Walls
Dry Stone Walling Association
Tel: 01539 567953
www.dswa.org.uk

Fencing
Jacksons Fencing
Tel: 01233 750740
www.jacksons-fencing.co.uk

Paving materials
Marshalls
Tel: 01422 312000
www.marshalls.co.uk

Paving expert
Tel: 01925 762 034
www.pavingexpert.com

RMC
Tel: 0870 2403030
www.rmc.co.uk

Ryburn Rubber
Tel: 01422 316323
www.ryburnrubber.co.uk

Retaining walls
For general advice read the Building
Research Establishment (BRE) booklet:
*Building Brickwork or Blockwork Retaining
Walls.* ISBN 1860811051

Building Regulations

Building Regulations Explanatory Booklet
Tel: 020 7944 4400
www.odpm.gov.uk/explanatory-booklet

Local Authority Building Control
Tel: 020 7641 8737
www.labc.co.uk

Northern Ireland Office
Tel: 020 90520 700
www.nio.gov.uk

Office of the Deputy Prime Minister
Tel: 020 7944 4400
www.odpm.gov.uk

Scottish Executive
Tel: 0131 556 8400
www.scotland.gov.uk

Construction methods

MODERN DEVELOPMENTS IN CONSTRUCTION METHODS
H+H Celcon
Tel: 01732 880520
www.celcon.co.uk

Kingspan TEK
Tel: 0870 850 8555
www.tek.kingspan.com/uk

FOUNDATIONS AND SERVICES
British Geological Survey
Tel: 0115 936 3143
www.bgs.ac.uk

CORGI
Tel: 0870 401 2200
www.corgi-gas-safety.com

The Environment Agency
Tel: 0870 08506506
www.environment-agency.gov.uk

Landmark Information Group
Tel: 020 7958 4999
www.landmarkinfo.co.uk

NICEIC
Tel: 0207564 2323
www.niceic.org.uk

WALLS
Brick suppliers
Hanson Bricks
Tel: 0870 525 8258
www.hanson.biz

Coatings and linings
British Gypsum
Tel: 0870 545 6123
www.british-gypsum.com

Knauf Drywall
Tel: 01795 424499
www.knauf.co.uk

Glass block suppliers
Luxcrete Ltd
Tel: 020 8965 7292
www.luxcrete.co.uk

Pittsburgh Corning (UK) Ltd
Tel: 0118 950 0655
www.pittsburghcorning.com

Shackerley (Holdings) Group Ltd
Tel: 01257 273114
www.shackerley.com

Karphosit clay blocks
Construction Resources
Tel: 020 7450 2211
www.constructionresources.co.uk

Ecomerchant
Tel: 01795 530130
www.ecomerchant.co.uk

Kingspan TEK
Kingspan Insulation Limited
Tel: 0870 850 8555
www.tek.kingspan.com/uk

LenoTec
Construction Resources
Tel: 020 7450 2211
www.constructionresources.com

Lindab
Lindab
Tel: 0121 585 2780
www.lindab.com

Masonry cavity walls/concrete blocks
The Aircrete Bureua
www.aircrete.co.uk

Brick Development Association
Tel: 01344 885651
www.brick.org.uk

Design and Materials
Tel: 01909 540123
www.designandmaterials.uk.com

H+H Celcon
Tel: 01732 886333
www.celcon.co.uk

Ibstock
Tel: 01530 261999
www.ibstock.co.uk

Tarmac Topblock Ltd for Durox
Tel: 01375 673344
www.topblock.co.uk

Thermalite
Tel: 08705 626500
www.thermalite.co.uk

Oak-frame house builders
Border Oak
Tel: 01568 708752
www.borderoak.com

Carpenter Oak Ltd
Tel: 01803 732900
www.carpenteroak.com

TJ Crump Oakwrights Ltd
Tel: 01432 353353
www.oakwrights.co.uk

Polystyrene block
Beco
Tel: 01724 747576
www.becowallform.co.uk

Insulating Concrete Formwork
Association
Tel: 0700 4266 273
www.icfinfo.org.uk

Styro Stone UK
Tel: 01580 767707
www.styrostone.com

Rendalath
BRC Building Products
Tel: 01785 22228
www.brc-building-products.co.uk

Softwood and stud frame
Timber Research and Development
Association (TRADA)
Tel: 01494 569601
www.asktrada.co.uk

UK Timber Frame Association
Tel: 020 7235 3364
www.timber-frame.org

Steko blocks
Construction Resources
Tel: 020 7450 2211
www.constructionresources.co.uk

Tile-backer boards
Aquapanel
www.aquapanel.co.uk

Wedi
www.wedi.co.uk

Timbatec stud framing
Construction Resources
Tel: 020 7450 2211
www.constructionresources.co.uk

Timber frame
Timber Research and Development
Association (TRADA)
Tel: 01494 569601
www.asktrada.co.uk

UK Timber Frame Association
Tel: 020 7235 3364
www.timber-frame.org

Timber-frame kit houses available from many companies including:
Build It magazine
For more timber-frame contact consult
Build It's Design and Packaging Company
Directory that it publishes each year.
Tel: 020 7772 8440
www.buildit-online.co.uk

Custom Homes
Tel: 01787 377388
www.customhomes.co.uk

Potton
Tel: 01480 401401
www.pottonhomes.com

FLOORS
Beam and concrete block/Hollowcore
Bison
Tel: 01283 495000
www.bison.co.uk

Hanson Building Products
Tel: 01773 602432
www.hanson.co.uk

Milbank Floors
Tel: 01787 223931
www.milbank.co.uk

The Precast Flooring Federation
The website is good for contacts and
basic information.
Tel: 0116 253 6161
www.pff.org.uk

RMC
Tel: 0117 937 3740
www.rmc.co.uk

Engineered timber
Eleco
Tel: 01252 334691
www.eleco.com/gang-nail/

Finnforest
Tel: 020 8420 0777
www.finnforest.co.uk

James Jones & Sons
Tel: 01309 671111
www.jji-joists.com

MiTek
Tel: 01384 451400
www.mii.com

TrusJoist
Tel: 0121 445 6666
www.trusjoist.com

Polystyrene infill blocks
Hanson Building Products
Tel: 01773 602432
www.hanson.co.uk

Timber joists
Timber Research and Development
Association (TRADA)
Tel: 01494 569601
www.asktrada.co.uk

ROOFS
Clay Roof Tile Council
Tel: 01782 744631
www.clayroof.co.uk

Redland
Tel: 0870 560 1000
www.redland.co.uk

The Tudor Roof Tile Company Ltd
(handmade tiles)
Tel: 01797 320202
www.tudorrooftiles.co.uk

Concrete tiles
Concrete Tile Manufacturers Association
Tel: 0116 253 6161
www.britishprecast.org

Marley Building Materials
Tel: 01675 468400
www.marleyroofing.co.uk

Redland
Tel: 0870 560 1000
www.redland.co.uk

Glass
Exitex
For capping systems
Tel: + 353 4293 71244
www.exitex.net

Glass and Glazing Federation
Tel: 0870 042 4255
www.ggf.org.uk

Pilkington
Tel: 01744 692000
www.pilkington.com

Solaglas
Tel: 024 7654 7400
www.saint-gobain-glass.com

Vitral
Tel: 01223 499000
www.vitral.co.uk

Green roof
Bauder
Tel: 01473 257671
www.erisco-bauder.co.uk

BHC
Tel: 01460 234582
www.greenroof.co.uk

Liquid waterproofing system (LWS)
European Liquid Roofing Association
Tel: 01444 417458
www.elra.org.uk

Kemper System
Tel: 01925 445532
www.kempersystem.co.uk

Mastic asphalt
Mastic Asphalt Council
Tel: 01424 814400
www.masticasphaltcouncil.co.uk

Metals
European Copper in Architecture Campaign
Tel: 01442 275700
www.cda.org.uk/arch

Lead Development Association
Tel: 020 7499 8422
www.lda.org

Lead Sheet Association
Tel: 01892 822773
www.leadsheetassociation.org.uk

Stainless Steel Advisory Service
Tel: 0114 267 1260
www.bssa.org.uk

Zinc Information Centre
Tel: 0121 362 1201
www.zincinfocentre.org

Single-ply roofing
Carlisle SynTec
Tel: 01235 848000
www.carlisle-syntec.com

Integrated Polymer Systems (UK) Ltd
Tel: 01969 625000
www.rubberfuse.co.uk

Single Ply Roofing Association
Tel: 0115 914 4445
www.spra.co.uk

Slate tiles
Alfred McAlpine
For Welsh slates.
Tel: 01248 600656
www.amslate.com

Thatch
National Society of Master Thatchers
www.nsmt.co.uk

Thatching Advisory Services Ltd
Tel: 01264 773820
www.thatchingadvisoryservices.co.uk

www.thatch.org

Three-ply polyester
Flat Roofing Alliance
Tel: 01444 440027
www.fra.org.uk

Timber shingles and shakes
John Brash Timber Importers
Tel: 01427 613858
www.johnbrash.co.uk

INSULATION
Cellulose fibre
Excel Industries
Tel: 01685 845200
www.excelsibre.com

Cork
Alumasc
Tel: 01744 648400
www.alumasc-exteriors.co.uk

Expanded polystyrene (EPS)
Vencel Resil
Tel: 020 8320 9100
www.vencel.co.uk

Extruded foamed polystyrene
Celotex
Tel: 01473 822093
www.celotex.co.uk

Dow Chemical Co Ltd
Tel: 020 8917 5049
www.dow.com/styrofoam

Kingspan
Tel: 01352 716100
www.insulation.kingspan.com

Foamed glass
Pittsburgh Corning (UK) Ltd
Tel: 0118 950 0655
www.foamglas.co.uk

Mineral fibre
British Gypsum
Tel: 0870 545 6123
www.british-gypsum.com

Eurisol – UK Mineral Wool Association
Tel: 020 7935 8532
www.eurisol.com

Knauf Insulation
Tel: 01795 424499
www.knauf.co.uk

Sheep's wool
Natural Building Technologies
Tel: 01844 338338
www.natural-building.co.uk

Second Nature UK Ltd
Tel: 01768 486285
www.secondnatureuk.com

Sprayed foam
Renotherm
Tel: 0800 169 3976
www.renotherm.com

Managing the Build

INSURANCE
ACE Europe
Tel: 0800 018 7660

BuildStore
Tel: 0870 870 9991
www.buildstore.co.uk

Conti Financial Servicese
Tel: 01273 772811
www.mortgageoverseas.com

DMS Services
Tel: 01909 591652
www.selfbuild.armor.co.uk

Zurich Insurance
Tel: 08000 966233
www.zurich.co.uk

PLANT/TOOL HIRE
Hewden
Tel: 0161 848 8621
www.hewden.co.uk

HSS Limited
Tel: 0845 728 2828
www.hss.co.uk

Jewsons Tel: 0800 539766
www.jewsons.com

Speedy Hire
Tel: 01942 720000
www.speedyhire.co.uk

Travis Perkins
Tel: 01604 752424
www.travisperkins.co.uk

SECONDHAND PLANT/EQUIPMENT
www.ebay.co.uk

WEATHER FORECASTS
BBC Weather
www.bbc.co.uk/weather

The Weather Channel
www.weather.co.uk

Home systems

HEATING
Boiler Efficiency Database
www.boilers.org.uk

Council of Registered Gas Installers
(CORGI)
Tel: 0870 4012200
www.corgi-gas-safety.com

The Energy Saving Trust
www.est.org.uk

Green Electricity Marketplace
Renewable energy availability
and tariffs
Tel: 01256 392701
www.greenelectricity.org

Heating and Hot Water Information
Council
Tel: 0845 600 2200
www.centralheating.co.uk

Institute of Domestic Heating and
Environmental Engineers
Tel: 01865 343 096
www.idhe.org.uk

National Energy Foundation
Tel: 01908 665555
www.natenergy.org.uk

Seasonal Efficiency of Domestic Boilers
in the UK (SEDBUK)
www.sedbuk.com

Approval of solid fuel
Hetas Ltd
www.hetas.co.uk

Oil-firing Technical Association
Tel: 0845 6585080
www.oftec.co.uk

Solid Fuel Association
Tel: 0845 6014406
www.solidfuel.co.uk

Boiler manufacturers
Baxi Potterton
Tel: 08706 060780
www.baxi.co.uk

Boulter Buderus
Tel: 01473 241555
www.boulter-buderus.co.uk

Grant UK
Tel: 0870 777 5553
www.grantuk.com

Keston Boilers
Tel: 020 8462 0262
www.keston.co.uk

Trianco Ltd
Tel: 0114 2572300
www.trianco.co.uk

Vaillant
Tel: 01634 292300
www.vaillant.co.uk

Worcester Bosch
Tel: 01905 752556
www.worcester-bosch.co.uk

Radiators
Beaumont
Tel: 01788 899100
www.beaumontcastiron.com

Caradon Stelrad
Tel: 0870 849 8056
www.stelrad.com

Clyde Combustion
Tel: 020 8391 2020
www.columnradiators.com

Feature Radiators
Tel: 01274 567789
www.featureradiators.co.uk

Ferroli Ltd
Tel: 0870 728 2882
www.ferroli.co.uk

The Radiator Company
Tel: 01342 302250
Scotland: 01236 780744
www.theradiatorcompany.co.uk

Underfloor heating
Devi
Tel: 01359 242400
www.devi.co.uk

NU-Heat
Tel: 01404 549770
www.nu-heat.co.uk

Robbens Systems
Tel: 01424 851111
www.underfloorheating.co.uk

VENTILATION
AllergyPlus
Tel: 0870 190 0022
www.allergyplus.co.uk

Building Services Research and
Information Association
Tel: 01344 465600
www.bsria.co.uk

Johnson and Starley
Tel: 01604 762881
www.johnsonandstarley.co.uk

Nuaire Group
Tel: 029 2088 5911
www.nuaire.co.uk

Silavent
Tel: 01252 878282
www.silavent.co.uk

Vent-Axia
Tel: 01293 526062
www.vent-axia.com

CENTRAL VACUUMING
SYSTEMS
BEAM/Electrolux
Tel: 01386 849000
www.thehhc.co.uk/beam.htm

Centravac
Tel: 0800 013 1517
www.centravac.co.uk

Elek Trends
Tel: 0800 083 0431
www.CVCdirect.co.uk

Villavent
Tel: 01993 778481
www.villavent.co.uk

HOME AUTOMATION
Automated Home
www.automatedhome.co.uk

Custom Electronic Design and
Installation Association (CEDIA)
Tel: 01462 627377
www.cedia.co.uk

SOUNDPROOFING
Building Research Establishment (BRE)
Tel: 01923 664300
www.bre.co.uk

InstaGoup
Tel: 0118 932 8811
www.instagroup.co.uk

Monarflex Acoustic Systems
Tel: 01727 830116
www.icopal.co.uk

Precast Flooring Federation
Tel: 0116 2536161
www.precastfloors.info

Prestressed Concrete Association
Tel: 0116 2536161
www.bridgebeams.org.uk

Soundcraft
Tel: 01959 533778
www.soundcraft-doors.co.uk

Timber Research & Development
Association (TRADA)
Tel: 01494 569601
www.asktrada.co.uk

SECURITY

Association of British Insurers (ABI)
Tel: 020 7600 3333
www.abi.org.uk

British Security Industry Association
(BSIA)
Tel: 01905 21464
www.bsia.co.uk

Electrical Contractors Association (ECA)
Tel: 020 7313 4800
www.eca.co.uk

Master Locksmiths Association (MLA)
Tel: 01327 262255
www.locksmiths.co.uk

National Neighbourhood Watch
Association
Tel: 020 7963 0160
www.neighbourhoodwatch.net

National Security Inspectorate (NSI)
Tel: 0870 2050000
www.nsi.org.uk

Secured by Design
A police initiative that will consult with
you about the best security measures for
your home
Tel: 020 7227 3423
www.securedbydesign.com

LIGHTING

Electrical Contractors Association (ECA)
Tel: 020 7313 4800
www.eca.co.uk

Institute of Electrical Engineers (IEE)
Tel: 020 7240 1871
For their *Wiring Regulations*:
www.iee.org/Publish/WireRegs

National Inspection Council for Electrical
Installation Contracting (NICEIC)
Has the largest registered members'
list of approved electrical installers.
Tel: 020 7564 2323
www.niceic.org.uk

Polaron
Tel: 01923 495 495
www.polaron.co.uk

Environmentally friendly self-build

Forests Forever
Tel: 017 1839 1891
www.forestsforever.org.uk

Forest Stewardship Council
Tel: 01686 413916
www.fsc-uk.info

Pan-European Forestry Certification
Council (PEFC), based in Luxembourg
Tel: +352 26 25 90 59
www.pefc.org

USEFUL CONTACTS

Association for Environment Conscious
Building (AECB)
Tel: 0845 456 9773
www.aecb.net

Building Research Establishment (BRE)
Tel: 01923 664000
www.bre.co.uk

The Centre for Alternative Technology
(CAT)
Tel: 01654 705950
www.cat.org.uk

Energy Saving Trust
Tel: 0845 727 7200
www.saveenergy.co.uk

Timber Research and Development
Association (TRADA)
Tel: 01494 569601
www.asktrada.co.uk

GREEN BUILDERS' MERCHANTS

Ecomerchant
Tel: 01795 530130
www.ecomerchant.co.uk

Green Building Store
Tel: 01484 854898
www.greenbuildingstore.co.uk

The Green Shop
Tel: 01452 770629
www.greenshop.co.uk

Mike Wye & Associates
Tel: 01409 281644
www.mikewye.co.uk

Natural Building Technologies
Tel: 01844 338338
www.natural-building.co.uk

SOLAR POWER

British Photovoltaic Association
Tel: 01908 442291
www.pv-uk.org.uk

Centre for Sustainable Energy
Tel: 0117 929 9950
www.cse.org.uk

Clear Skies
Tel: 0870 243 0930
www.clear-skies.org

Environ
Tel: 0116 222 0222
www.environ.org.uk

Solar Trade Association
Tel: 01908 442290
www.greenenergy.org.uk/sta

WIND TURBINES

British Wind Energy Association
(BWEA)
Tel: 020 7689 1960
www.bwea.com

Clear Skies
Tel: 0870 243 0930
www.clear-skies.org

Good Energy
Tel: 0845 4561640
www.unit-e.co.uk

Proven Energy
Tel: 01560 485570
www.provenenergy.com

EARTH-SHELTERED HOUSING

British Earth Sheltering Association
(BESA)
Tel: 01600 860359
www.besa-uk.org

Hockerton Housing Project
Tel: 01636 816902
www.hockerton.demon.co.uk

The Underground House
www.theundergroundhouse.org.uk

STRAW-BALE HOUSES

Amazon Nails
Straw bale building, training and
consultancy
Tel: 0845 458 2173
www.strawbalefutures.org.uk

The Straw Bale Building Association
Tel: 01442 825421
www.strawbalebuildingassociation.org.uk

www.strawbale-building.co.uk

Where to live

www.themovechannel.com

CARAVANS
Acorn Caravan and Trailer Hire
Tel: 0118 971 2918
www.acorncaravans.co.uk

The Caravan Club
Tel: 01342 326944
www.caravanclub.co.uk

The Caravan Company
Tel: 01908 586000
www.thecaravancompany.co.uk

REMOVALS
Bishops Move Group
Tel: 020 7498 0300
www.bishops-move.co.uk

Britannia
Tel: 0800 622535
www.britannia-movers.co.uk

British Association of Removers
Tel: 020 8861 3331
www.removers.org.uk

www.helpiammoving.com

www.houseremovals.com
Tel: 0800 980 5866

Moves
Tel: 08700 104410
www.moves.co.uk

Pickfords
Tel: 0800 289229
www.pickfords.co.uk

reallymoving.com
Tel: 0870 870 4851
www.reallymoving.com

STORAGE
The Big Yellow Self Storage
Company
Tel: 0800 783 4949
www.thebigyellow.co.uk

Bishops Move Group
Tel: 020 7498 0300
www.bishops-move.co.uk

Pickfords Self Store
Tel: 0800 212102
www.pickfordsselfstore.co.uk

Safestore
Tel: 0800 444800
www.safestore.co.uk

Index

disability access 101
energy conservation 96, 97
environmentally-friendly options 188
garages 88
height 61
insulation 187, 188
security 180
soundproofing 179
styles 79
dormer windows 50, 80, 239
double glazing 188
downlighters 184
drains and drainage 225–6
air leakage tests 95
Building Regulations 95
crossing foundations 237
foul-water system 111
inspection chambers 95
paving 86
rainwater 111
drawings, outline planning permission 228
driven pile foundations 125
driveways
lighting 182
planning permission 86–7
subbases 85
trees and 239–40
dry-stone walls 236
dust mites 174, 175
DVDs, home automation systems 177

E

earth-sheltered houses 192–4
earth walls 114
eating areas, planning 66, 69
edgings, paving 86
Edwardian-style houses 82
cornicing 81
doors 79
staircases 81
windows 80
Electrical Contractors Association (ECA) 182
electrical sockets
disability access 101
fitting 183
height 239
electricians 156, 182–3
electricity
caravans 197
connecting to services 110, 111
costs 235

lighting 182–4
safety 183
solar power 110, 188–91
temporary supplies 170
wind turbines 110, 191–2
see also services
elevations, front 78
employer's liability insurance 22
energy conservation
boilers 172
Building Regulations 96–100
earth-sheltered houses 192
insulation 174
solar power 188–91
ventilation systems 175
wind turbines 191–2
engineered timber joists, suspended floors 135, 138
engineers, structural 155
entertainment, automated systems 176–7
Environment Agency 34, 45, 110, 111
environmentally-friendly options 186–95
brick-and-block houses 186–8
doors 188
earth-sheltered houses 192–4
external walls 114
green roofs 146
ground floors 116
heating systems 110
insulation 120, 188
sewage disposal 111
solar power 188–91
straw-bale buildings 194–5
timber-frame houses 187–8
wind turbines 191–2
windows 188
equipment
hiring plant 166–9
protective 165
estate agents
buying sites 53, 54
fees 25
finding sites 28, 29, 30
valuing sites 55
estimates 12–15
builders' 233
hiring scaffolding 169
written estimates 161
Euro mortgages 226

excavations
foundations 94
mini excavators 166
safety 165
existing use, and planning permission 41
expanded polystyrene insulation 148
extensions
Building Regulations 91
maximizing conversions 38
permitted development rights 48, 49, 231
timber-frame houses 236
VAT (value-added tax) 19
extensive green roofs 146
exterior lights 182
external walls
cladding 113, 127–8
construction methods 112–14, 127–30
extruded foam polystyrene insulation 148

F

fees
architects' 71–2, 232–3, 241
Building Regulations 93
independent project managers 153, 158
Land Registry 25, 26
outline planning permission 42
planning permission 43
purchase costs 24–6
fences 83
acoustic fences 239
permitted development rights 49
site security 164, 180
types of 84–5
fibre glass insulating board 120
fibrous insulation 119–20, 147
finance see budget and finance
fire safety
earth-sheltered houses 193
straw-bale buildings 194
timber-frame houses 113
first-aid box 165
first-fix 157
electricity 183
first-floor plans 77
fixtures and fittings, design 79–82

flat raft foundations 123
flat roofs 118–19, 145–6
flats, soundproofing 239
flood-risk areas 34–5
floodlights 184
floor lights 184
floor sanders 166
floors
concrete 103
construction methods 115–17, 133–40
insulation 119–20, 133
soundproofing 178–9
wet rooms 117
foam insulation 149
foamed glass, insulation 148
foils, insulation 149
footpaths 32
footwear, protective 165
Forestry Stewardship Council (FSC) 188
formal tender, buying sites 53
foundations 107–9, 157
Building Regulations 94–5
concreting 94, 95
construction methods 122–6
costs 234
deep foundations 125–6
depth 108
drains crossing 237
excavation 94
garages 88
garden walls 83
ground surveys 107–8
retaining walls 239
shallow foundations 122–4
straw-bale buildings 194, 195
trees and 237
underpinning 234
front elevations 78
fuels
choosing 172–3, 238
costs 174
furniture
on architects' plans 76
bedrooms 62
kitchens 68
moving house 198
planning rooms 61
reception rooms 66
storage 197–8

G

garages 87–8, 234
garden buildings 86
permitted development rights 48–50

gardens
building on 52–3
lighting 182
security 180
selling plots 227, 231
size 224
gas
boilers 172, 174, 238
caravans 197
connecting to services 110
see also services
gates
permitted development rights 49
site security 164
gazebos 86
gazumping 55
generators 170, 191
Georgian-style houses 82
cornicing 81
doors 79
staircases 81
windows 80
glass
energy conservation 100
insulation 148
internal walls 132
roofs 142, 146
see also windows
glass blocks 132
glass fibre insulation 147, 188
glazing see glass
Gould, Andrew and Karen 204–6
graffiti 85
grants
solar panels 191
wind turbines 192
gravel 85
green belts
barn conversions 229–30
permitted development rights 31, 51
'green' building see environmentally-friendly options
green roofs 146
ground-bearing floors 115–16, 133–4
ground-floor plans 76
ground floors
Building Regulations 95
construction methods 115–16, 133–7
insulation 116, 133
ground surveys 33–4, 107–8
groundworkers 155

site surveys 35
suspended floors 115–16, 134–7
swimming pools 49
switches see light switches

T

tanks, cold water 173
task lighting 184
tax
    capital gains tax (CGT) 23–4, 226, 227
    council tax 227
    stamp duty land tax 25, 26
    VAT (value-added tax) 16–19, 161, 226
technologists 70, 71, 155
    drawings 228
    fees 72
    finding 73
    as project managers 154
Tek Building System 106–7, 114, 128–9
telephones
    connecting to services 110, 111
    costs 235
    monitored alarm systems 181
    see also services
television, closed-circuit 176, 181
temperature, earth-sheltered houses 192
temporary accommodation 196–7
tenders
    buying sites 53–4
    private sales 224
    sending proposals out to tender 15
terraced houses, permitted development rights 48, 49
terraces, roof 119
thatched roofs 144
thermal mass 194
thin-joint mortaring 103
tidiness, on site 163–4
tiles
    roofs 119, 141
    tile-hung walls 113
    tilers 156
Timbatec stud framing 115, 132
timber
    cladding 113
    fences 84–5
    log walls 130

roof shingles and shakes 144
soundproofing wooden floors 179
suspended floors 134–5, 138
upper floors 117
windows 188
timber-frame houses
    building costs 11
    case studies 201–3, 204–6, 210
    construction methods 103–4, 113, 127–8, 129, 131
    energy conservation 96
    environmentally-friendly options 187–8
    extensions 236
    insulation 119–20, 187
    oak frames 129
    post and beam construction 105, 201–3
time
    anticipating and preventing delays 159
    causes of delays 159
    completion dates 158, 160
    overrunning 160
    slack timekeeping 160
toeboards, scaffolding 169
toilets
    disability access 101
    on site 165–6
    water conservation 97
tools, hiring 166–8
topsoil 108
towers, scaffold 166
towns, brownfield sites 52–3, 54
transformers, low-voltage lighting 184
trees
    assessing sites 33
    and driveways 239–40
    and foundations 109
    tree preservation orders (TPOs) 52, 237
trellis 85
trench blocks 237
trench fill foundations 122
trenches
    installing services 235
    see also foundations
trestles, scaffold 166
triple-glazed windows 188, 239
trusses, roof 238–9

Tudor-style houses 82
    lintels 81
    windows 80
turbines, wind 110, 191–2

U

U-values, heat loss 96, 98, 101
underfloor heating 235
underground houses 56, 192–4
underground streams 235–6
underlay, roofs 119
underpinning foundations 234
uplighters 184
upper floors
    construction methods 116–17, 138–40
    finishes 117
urban areas, brownfield sites 52–3, 54
utility companies
    finding plots 28, 30
    right of entry 32
    see also services
utility rooms, planning 68
Uzzell, Charles 212

V

vacuuming, central systems 175–6
valuations
    for council tax 227
    house prices 9
    sites 55, 224
vapour control
    ground floors 116
    timber-frame houses 187–8
VAT (value-added tax) 16–19, 161, 226
ventilation 174–5
    earth-sheltered houses 192, 193
    ground floors 116
vibro compaction foundations 126
vibro replacement foundations 126
villages, infill sites 52

W

Wales, planning applications 46, 47
wall lights 184
wallboard 113
walls
    on architects' plans 76

brick-and-block construction 102–3
cavity walls 96–7, 112–13, 127, 238
cladding 50, 113, 127–8
coatings 113, 114–15
construction methods 111–15, 127–32
dry-stone walls 236
earth-sheltered houses 193
energy conservation 96–7
garages 88
garden walls 83–4, 85
insulation 119–20
linings 113, 114
party-fence walls 83
permitted development rights 49
retaining walls 239
and security 180
soundproofing 114, 178–9
straw-bale buildings 194–5
timber-frame construction 103–4, 113, 127–8, 131
wattle and daub 194
wiring 183
wardrobes 62
warranties, structural 19–21
waste disposal 163–4
water
    boreholes 110, 240
    for caravans 197
    combination ('combi') boilers 173
    connecting to services 110, 111
    conservation requirements 97
    costs 235
    drainage systems 111
    and foundations 109
    heating systems 100
    solar heating 173, 189, 190, 191
    tanks 173
    temporary supplies 170
    underground streams 235–6
    wet rooms 117
    see also services
water pressure 173
waterproofing
    earth-sheltered houses 193

straw-bale buildings 194, 195
wattle and daub 194
weather 159
websites, finding plots 30
welfare, on site 165–6
wet rooms 117
wheelchair access 87, 101
wide strip foundations 122
willow fences 85
wind turbines 110, 191–2
windows
    aluminium 188
    on architects' plans 76, 78
    architraves 81
    dormer windows 50, 80, 239
    double glazing 188
    earth-sheltered houses 192
    energy conservation 96, 97
    environmentally-friendly options 188
    installers 156
    insulation 187
    locks 180
    permitted development rights 50
    plastic 188
    roof lights 238
    styles 80
    timber 188
    triple glazing 188, 239
wireless networks, home automation systems 177
wiring
    home automation systems 177–8
    lighting 183
withholding money 160
wood see timber
wood-burning appliances 110
wood pellet boilers 173
wooden doors, security 180
wool insulation 120, 147, 188

Z

zinc roofs 143, 145
Zurich Municipal 20, 21

About the contributors

Jane Crittenden specializes in writing about house-building and renovating, commercial building and construction, interiors and architecture. She used to be deputy editor on Build It and now writes for magazines in the UK, Australia and New Zealand including *Build It*, *Houses*, *Your Home & Garden*, *Home & Entertaining*, *Urbis*, *Architecture* and *Progressive Building*.

Mike Dade and **Roy Speer** are contributing editors to *Build It* who first wrote for the magazine in 1992. They are both Chartered Planning and Development Surveyors, with over 36 years of professional experience in the property industry between them. Both Roy and Mike have degrees in Estate Management and are partners in their own planning consultancy practice, Speer Dade Planning Consultants, founded in 1991. They have written a number of books on land finding and planning and are regular speakers on these subjects. Roy lectures on planning at Kingston University both to students and property professionals.

Nigel Grace is a chartered quantity surveyor MRICS and an experienced self-builder. His most recent build, completed in 2003, was a four-bedroom family home that featured in Build It and is currently building two more houses. He has worked in the building industry for 18 years and written for *Build It* since 1999 while also contributing regularly to RICS magazine, *Building* and *The Financial Times*.

Catherine Monk edits *Build It*, the magazine for people who build their own home or renovate properties, and is the *Sunday Times* self-build expert. She is an award-winning editor who regularly speaks about the self-build industry on TV and the radio. Catherine's enthusiasm for property extends to renovating houses and DIY.

Laura Pank is a regular contributor for *Build It*. Her experience of self-build started when she moved into an arts and crafts house that needed extensive renovation and conversion. Laura contributed to the design of the extension which rewarded her with a crash course induction in all aspects of self-build from planning, materials and construction methods.

Kevin Smith is the former deputy editor of *Build It*. Along with a first class degree in Journalism from the University of Manchester, Kevin has over two years experience writing about self-build.

Peter Smithdale MSc ARCH RIBA is an architect for Constructive Individuals. For over ten years Peter has worked and with volunteers to design and build several community and residential buildings. One such project involved erecting a timber-frame house in just three weeks in 2000. He is a regular contributor for *Build It*.

Build It is a monthly magazine for people who build their own home, or who renovate or convert properties. www.buildit-online.co.uk , 020 7772 8440.

Acknowledgements

All images © *Build It Magazine*

Apart from the following:

Anderson Windows (tel: 01283 511122, www.andersonwindows.com) 10 bottom left, 50 bottom left.

Border Oak (tel: 01568 708752, www.borderoak.com) 7 top, 105 bottom left.

Bradstone (tel: 0800 975 9828, www.bradstone.com/garden) 84 top right, 84 bottom.

H + H Celcon (tel: 01732 886333, www.celcon.co.uk) 22 top left, 41 top left, 105 bottom right, 106 bottom left, 181 bottom.

Construction Photography 20 bottom left, 93 bottom right, 109 bottom, 153 centre right.

Copper Development Association (tel: 01442 275 700, www.cda.org.uk) 118 bottom right.

Corbis UK Limited 2; C/B Productions 70 top right; /Bill Varie 24 bottom.

Federation of Master Builders (tel: 020 7242 7583, www.fmb.org.uk) 152 centre left, 154 top right, 164 top left.

Glass Houses (tel: 020 7607 6071, www.glass-houses.com) 92, 99.

Kingspan Insulation (tel: 01544 388601, www.insulation.kingspan.com) 107 top left.

Lassco (www.lassco.co.uk) 41 bottom right.

Nexus Radiators (from **Parker Hobart PR**, tel: 020 7584 1744) 100 top left.

Octopus Publishing Group Limited 4-5; /Sebastian Hedgecoe, Roger Black Development 63.

Papa Architects Limited (tel: 020 8348 8411) 60 bottom right, 60 bottom left, 67 bottom.

Parker Hobart Associates PR (tel: 020 7584 1744) 69 top, 69 bottom right.

Photolibrary.com 119 top right, 196 top, 198 bottom left.

Potton (tel: 01480 401401, www.potton.co.uk) 16 bottom left, 58 bottom right, 69 bottom left, 90 bottom left, 102 top, 104 bottom, 114 bottom right, 117 bottom, 150 bottom left, 158 bottom, 166 top right, 169 bottom right.

Richard Burbidge (tel: 01691 678201, www.richardburbidge.co.uk) 80 bottom, 81 bottom right, 81 bottom left.

Science Photo Library/Simon Fraser 34 bottom; /Alan Sirulnikoff 195 bottom.

Veissman (tel: 01952 675 000, www.viessmann.com) 172 centre right, 186 bottom left, 188 bottom right, 190 bottom.

Welsh Oak Frame (tel: 01686 688000, www.welshoakframe.com) 58 centre right, 62 bottom, 97 Bottom.

Wickes Building Supplies Limited (tel: 0870 6089001, www.wickes.co.uk) 64 bottom, 179 bottom right.

Executive Editor: Katy Denny

Managing Editor: Clare Churly

Editor: Anna Southgate

Executive Art Editor: Geoff Fennell

Designer: Grade Design Consultants

Photographers: Bob Atkins, Darren Chung, Tony Hall, Nigel Rigden

Picture Library Manager: Jennifer Veall

Picture Researcher: Zoe Willows

Production Manager: Ian Paton